Tree Growth and Environmental Stresses

THEODORE T. KOZLOWSKI

UNIVERSITY OF WASHINGTON PRESS

Seattle and London

Copyright © 1979 by the University of Washington Press
Printed in the United States of America

All rights reserved. No part of this publication may be reproduced or transmitted in any form or by any means, electronic or mechanical, including photocopy, recording, or any information storage or retrieval system, without permission in writing from the publisher.

Library of Congress Cataloging in Publication Data
Kozlowski, Theodore Thomas, 1917–
   Tree growth and environmental stresses.

  (The Geo. S. Long publication series)
  Bibliography: p.
  1. Trees—Growth—Addresses, essays, lectures.
2. Forest ecology—Addresses, essays, lectures.
I. Title. II. Series.
SD396.K66   582'.16'03   78–10815
ISBN 0–295–95636–4

Theodore T. Kozlowski is A. J. Riker professor of forestry and director of the Biotron at the University of Wisconsin, Madison. He is the author of many books and articles and has been the recipient of a number of awards in the field of forest research.

THE GEO. S. LONG PUBLICATION SERIES

THE GEO. S. LONG PUBLICATION SERIES

American Forest Policy in Development by Stephen H. Spurr

Man, Land, and the Forest Environment by Marion Clawson

Renewable Resource Management for Forestry and Agriculture by James S. Bethel and Martin A. Massengale

Tree Growth and Environmental Stresses by Theodore T. Kozlowski

THE GEO. S. LONG PUBLICATION SERIES

The Geo. S. Long Fund was established in 1975 to promote a better understanding of forestry, natural resources, and conservation. The endowment was made by Miss Helen Long of Tacoma to the University of Washington in memory of her father, a distinguished timber industry executive and natural resources conservationist, who opened the first Weyerhaeuser office in Tacoma in 1900 and served as the firm's manager, vice president, and chairman of the executive committee.

The generous financial support provides an income to the University of Washington to carry out the intent of the founder with provisions that allow the dean of the College of Forest Resources to administer a program of lectures, publications, travel, and library acquisition.

The establishment of the lecture series with the introduction of distinguished authorities to the academic and forestry community has been endorsed by Miss Long as a principal effort of the fund. Publication of the Long Lectures as well as the work of other educators and scientists in associated disciplines will supplement and extend the influence of the program.

The terms of the gift are broad enough to allow the university to pursue scholarly excellence in such areas as special collections and the attendance of faculty at academic and professional meetings of special merit.

Seventy-five years elapsed between the time of Geo. S. Long's first arrival in the Pacific Northwest and the establishment of this series. With Miss Long's authorization and support the university will continue the legacy to work for a better understanding of forestry, natural resources, and conservation through academic efforts, a distinguished lecture series, and these publications.

Contents

INTRODUCTION	3
COMPLEXITY OF ENVIRONMENTAL STRESSES AND TREE RESPONSES	9
WATER SUPPLY AND TREE GROWTH	56
THE ENVIRONMENTAL IMPACT ON SEEDS AND SEEDLINGS	108
CHALLENGES OF RESEARCH IN FOREST BIOLOGY	158
APPENDIX 1: HERBICIDES MENTIONED	169
APPENDIX 2: SCIENTIFIC AND COMMON NAMES OF SPECIES MENTIONED	170
LITERATURE CITED	175

TREE GROWTH AND ENVIRONMENTAL STRESSES

Introduction

WHEN DEAN BETHEL INVITED ME TO PRESENT THE Geo. S. Long Lectures at the College of Forest Resources I accepted with enthusiasm for a number of reasons. I interpreted his invitation as an opportunity to set my own experiences against cognate research in environmental biology and to talk freely about the impacts of environmental stresses on forest ecosystems, something with which I have been concerned for some thirty years. Another reason was to call attention to the need for greater recognition of the fact that our own technology is, sometimes inadvertently, superimposing some very dangerous environmental impacts on the stresses already bestowed by nature on forest trees. I also saw this occasion as an opportunity to share with students my enthusiasm for the challenges available for truly exciting careers in environmental biology as it pertains to forest ecosystems. Yet another reason for accepting this assignment was that I have been impressed for many years by the concern that the faculty and graduates of the College of Forest Resources have demonstrated with the impact of adverse environments on forest trees.

I suppose I first became concerned with effects of stresses on trees a very long time ago, in fact between my junior and senior years in college, when I worked with a Dutch elm disease crew and saw thousands of trees decimated by a biotic stress that involved a lowly fungus. Shortly thereafter I saw the effects of the 1938 hurricane that uprooted vast acreages of forest trees in the New England states. You will recall that years were required to salvage the tremendous amount of timber blown down by that hurricane. Since that time I have had many opportunities to see at first hand the destruction that periodic environmental extremes can cause in both temperate and tropical forests. Yet the violence that environmental extremes do periodically to forest ecosystems should not obscure the perhaps even more important and subtle effects of sustained environmental stresses in causing tremendous growth losses without inducing visible injury to trees.

Despite the harm that unfavorable environments do to forest ecosystems, we see much evidence that some forest trees are uniquely resistant to environmental stress. Certain trees are the world's oldest living things, having survived for thousands of years, often in an extremely hostile environment. We also recognize that survival of these trees requires integration and coordination of physiological processes occurring in widely separated roots and shoots. It strikes me as remarkable that trees can live for more than 3,000 years and maintain the necessary transport of food, water, hormonal growth regulators, and minerals over distances of several hundred feet. Furthermore the lengths of these translocation paths increase progressively. The survival of old and large trees is even more remarkable when we consider that the stem tissue through which carbohydrates move between the crown and roots is a layer of inner bark that is only a fraction of a millimeter thick. It thus seems obvious that from a physiological viewpoint, some trees must be beautifully coordinated throughout their lives in order to survive.

Wood is a renewable resource that will become increasingly important to mankind as fossil hydrocarbons are gradually exhausted. In the temperate zone, wood is being renewed at a rate far below the potentially attainable rate. Trees grow slowly and the rate at which wood will be renewed will depend on our understanding of the biological processes involved in tree growth, the ecological nexus that controls those processes, and our abilities to alter and manipulate the environment so as to maintain conditions for physiological processes of forest trees to be favorably coordinated throughout their development. At the same time, however, we must be very careful that we do not damage the stability of the forest ecosystem and thereby prevent the renewability of the product we harvest.

There will be heavy overtones in these lectures of the importance of various plant processes that are the critical intermediaries through which heredity and environment interact to influence growth. This purposeful emphasis is based on Kramer's (1956) suggestion that foresters can learn how to grow trees by learning more about how trees grow. Kramer meant by this that the potential contribution of physiology to forestry lies in identifying and characterizing the fundamental processes of trees which influence their growth, and demonstrating how those processes are affected by heredity and environment. The environmental changes which alter tree growth do not do so directly but rather indirectly through their influence on rates and balances among photosynthesis, respiration, assimilation, hormone synthesis, absorption of water and minerals, translocation of growth requirements (carbohydrates, hormones, water, and minerals), and other processes and physicochemical conditions. Hence, the important forestry problems of seed production, seed germination, wood production, wood quality, and seed and bud dormancy all involve physiological balances and adjustments. Furthermore, growth inhibition or death of trees in response to environmental stresses are preceded by a series of abnormal physiological events. The changes in internal processes of trees are close to the event, growth response, and they alter it.

The first lecture, "The Complexity of Environmental Stresses and Tree Responses," is an overview of characteristics of both natural and man-induced environmental stresses on forest ecosystems. It outlines physiological requirements for growth of unstressed trees and interactive effects of environmental factors on physiological processes and growth. It also addresses some of the problems involved in analyzing the contributions of individual environmental factors to tree response. Particular attention is given to effects of light, temperature, mineral supply, and environmental pollution.

The second lecture, "Water Supply and Tree Growth," analyzes the impact of both water deficiency and excess on forest trees. Attention is given to the importance and sources of water, the creation and quantification of internal water deficits, physiological and growth responses of various tissues and organs as water stresses develop, and adaptations for drought resistance. The sequential physiological dysfunctions and growth responses of trees to flooding are also summarized.

The third and final lecture, "The Environmental Impact on Seeds and Seedlings," provides an assessment of environmental stresses on trees in the stages of their greatest mortality risk. Physiological requirements are outlined for seed germination and early seedling development. Primary attention is given to the physiological role of cotyledons in seedling development. Extreme sensitivity of seedlings in the cotyledon stage to environmental stresses and to naturally occurring and applied biocides is emphasized. The importance of physiological vigor of planting stock is also discussed. Finally some consideration is given to research needs and rationale for coping with the impact of environmental stresses on forest ecosystems.

Special thanks are extended to Dean Bethel for his leadership in fostering the Geo. S. Long Lecture Series, and for giving me the opportunity to discuss problems of mutual interest with students and faculty. I am deeply grateful for his confidence in not placing any constraints on what these lectures would encompass. I also express sincere appreciation to Miss Helen Long for her foresight, good judgment, and concern with our renewable natural resources as shown by establishment of this lecture series to honor her very distinguished father.

Complexity of Environmental Stresses and Tree Responses

THE WAYS IN WHICH TREES RESPOND TO ENVIRONmental stresses have concerned man for centuries. The most obvious responses are drastic ones, such as injury or death of trees, but perhaps even more important are the many less obvious and subtle effects that lead to growth inhibition.

Climatic restrictions on boundaries of tree ranges are well documented. The amount and seasonal distribution of rain, extent and timing of freezing, and amount of summer heat can contribute to stress in plants. Unfavorable climate is an effective barrier to plant migration only when all the habitats within a region lie beyond the ecological amplitude of a species. This is shown by the linear extensions of the mesophytic species, *Juglans nigra* and *Carya cordiformis*, into the mid-continental steppe where water supplies are adequate, as in ravine bottoms (Daubenmire 1978).

When the boundaries of arctic air change, the limits of the boreal forest in North America also change. Records of these shifts in species migration may be found in the podzolic soils developed under these forests. North of the present boreal forest for 1,000 miles east to west, podzolic soils alternate with grey-brown soil of the treeless tundra, providing a record of shifting of the boreal forest boundary with expansion and contraction of the arctic air environment (Bryson and Murray 1977). In the forest-tundra ecotone region of the Washington and Oregon Cascades, invasion of gymnosperms, especially *Abies lasiocarpa* and *Tsuga mertensiana,* into subalpine meadows occurred during 1928–37. This migration was attributed largely to a period of warm climate and long snow-free seasons. Snow pack, through its influence on length of growing season, is one of the most important factors determining the position of forest-meadow ecotones in subalpine zones (Franklin et al. 1971).

Environmental stresses adversely affect trees in different ways. They may induce a direct plastic strain, recognizable by rapid appearance of injury. An example is the killing of physiologically active plants by sudden exposure to freezing temperatures. Environmental stresses may also produce a noninjurious, reversible, elastic strain which, if maintained for a long enough time, may induce an irreversible and injurious plastic strain. Additionally an environmental strain may cause injury by inducing a secondary stress. For example, high temperature may induce plant water deficits, which cause injury. Such secondary stress injury may not develop for a considerable time, hence long exposure to the primary stress may be necessary. Conceivably a secondary stress may induce a tertiary stress that also may cause injury or growth loss.

Levitt (1972) classifies environmental stresses as either biotic or physicochemical: the former encompassing infection or competition by other organisms; the latter including effects of radiation, water, temperature, chemical substances, wind, pressure, sound, etc. (fig. 1).

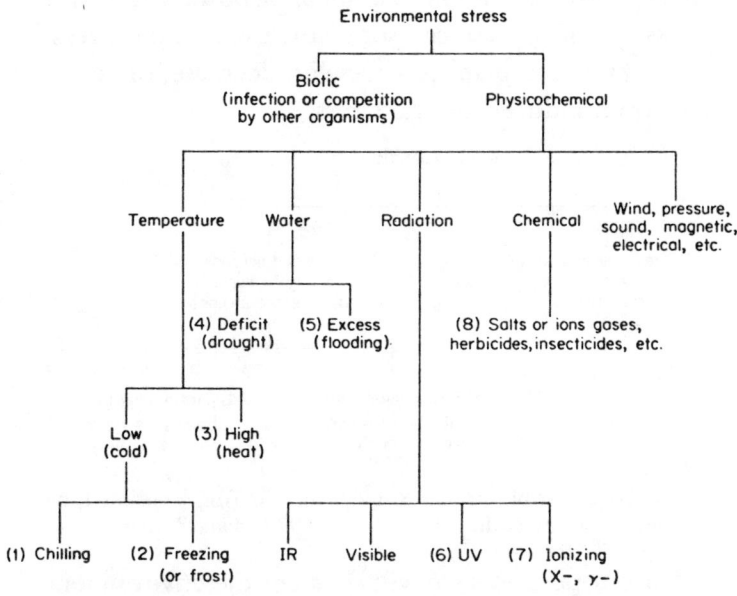

Fig. 1. Important environmental stresses affecting tree growth. Considerable data are available on resistance to the numbered stresses. Reprinted, by permission of the author, from Levitt 1972, diag. 2.2.

Fortunately, trees, like other organisms, can adapt to certain stresses. When stressed they gradually change to decrease or prevent strain. Adaptations that arose by evolution over a long time are stable; or they may be unstable depending on the state of plant development and the environmental factors involved in the stress situation.

As may be seen in Figure 2 a number of environmental stresses give rise to various degrees of resistance adaptations in plants. Stress resistance may reflect stress avoidance, stress tolerance, or both. Whereas a stress avoiding plant can somehow exclude the stress, a stress tolerant plant can prevent, decrease, or repair the strain induced by the stress.

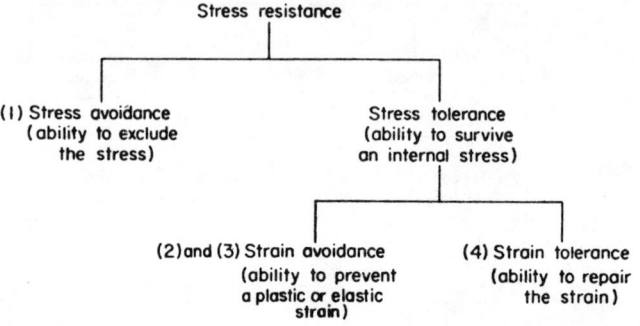

Fig. 2. Four possible stress resistance mechanisms. Reprinted, by permission of the author, from Levitt 1972, diag. 2.4.

For our purposes we will consider the environment to be the sum of all external forces affecting tree growth. Inasmuch as growth is an integrated response to physiological changes regulated by a complex of many fluctuating and interacting environmental factors it is difficult to evaluate the contribution of a single stress factor to growth. Some of the change in the environment of a tree is cyclic and some is continuous. Furthermore growth characteristics vary between and within tropical and temperate zone species as well as clones and cultivars. Growth characteristics also vary markedly in different parts of a tree and they vary with age of trees. The effects of an environmental stress on trees often depend on the phenological stage and physiological status of the tree at the time of occurrence of the stress.

Another problem is that the relative importance of environmental factors in growth control changes during the growing season. For example, in Wisconsin, correlation of cambial growth of *Quercus ellipsoidalis* with temperature decreased toward late summer as soil moisture was progressively depleted and growth was limited by internal water deficits (Kozlowski, Winget, and Torrie 1962). Still another difficulty involves distinguishing between correlations and cause and effect relations of environmental factors and growth. High correlation between changes in an environmental factor and tree growth may suggest that the factor is significantly influencing growth controlling processes. However, this factor may only be correlated with some other factor that is regulating growth processes but is not even included in the analysis.

The effect of an environmental stress may not be evident for a very long time. Lag responses of tree growth to environmental changes are well illustrated by effects of stand thinning on cambial growth. After thinning, the released trees respond to more favorable environmental conditions by slowing down upward crown recession and increasing crown width and leaf volume. These changes lead to increased production and basipetal flow of carbohydrates and hormonal growth regulators. Eventually a more tapered stem is produced by greater stimulation of xylem production toward the stem base than at upper stem levels or by redistributing xylem increment to favor the lower stem. Often, however, a significant increase in cambial growth may not be evident in the lower stem until a few years after the thinning.

Variations in lag responses are also shown in effects of environmental stress on shoot growth. There are three basic patterns of shoot growth in forest trees. In one group of species the shoots are fully predetermined in the winter bud—that is, the shoots exhibit "fixed" growth. Shoot formation involves bud differentiation during one year and extension of the preformed parts within the bud into a shoot during the next year. Northern pines and some maples and ashes are of this type. A second group of species exhibits "free" growth. This involves elongation of a shoot by simultaneous initiation and elongation of new stem units as well as expansion of preformed parts. Shoots of species that exhibit free growth often have two sets of leaves: (1) early leaves that were relatively well developed in the winter bud, and (2) late leaves which continued to form and elongate while the shoot internode was expanding. Woody plants that exhibit free growth include *Populus, Malus, Betula, Eucalyptus,* and *Larix.* In a third group, such as the southern pines and many species of tropical angiosperms, shoot elongation consists of a series of recurrent growth flushes from opening of a series of buds produced during the same growing season. When recurrently flushing species are moved to the hot and wet tropics their shoots may grow continuously throughout the year rather than in periodic surges (Kozlowski and Greathouse 1970).

In woody plants with fixed growth, potential shoot growth is largely governed by the number of stem units present in the unopened bud. Hence the phase of bud formation determines ultimate shoot length. No matter how favorable the environment is during the year of bud expansion, shoots of species with fixed growth expand for only a few weeks in the early part of the frost-free season. Thus, environmental conditions during the previous year, when buds formed, control ultimate shoot length of such species more than those during the year of bud expansion into a shoot. A favorable environment during the year of bud formation causes formation of large buds, which produce long shoots in the subsequent growing season. Bud size can thus be used as an index of shoot growth potential of certain species and their provenances and of different shoots on the same tree. For example, in *Pinus resinosa* the amount and rate of shoot growth is greater in upper whorls, which have the largest buds, than in lower whorls (Kozlowski, Torrie, and Marshall 1973).

Several investigators found strong correlations between shoot growth of species with fixed growth and weather conditions of the previous year. For example, when late-summer temperatures were low, the period of bud differentiation of *Picea abies* plants was short and this was reflected in production of short internodes during the following year (Heide 1974). Mikola (1962) emphasized that annual height growth of *Pinus silvestris* in Finland was determined largely by air temperature of the previous summer. Clements (1970) noted that irrigation of *Pinus resinosa* trees in late summer caused formation of large buds, which produced long shoots in the subsequent growing season. By comparison, late summer droughts induced small buds, which produced shoots with short internodes. Irrigation in the spring did not appreciably influence shoot length during the same year, emphasizing the importance of shoot predetermination on shoot length.

The dependence of shoot length on prior-year weather applies primarily to species with preformed shoots whose internodes complete expansion during the very early part of the frost-free season. Thus, whereas environmental stresses late in the summer often do not limit current-year shoot expansion of species with predetermined shoots (fixed growth) they will inhibit the expansion of shoots of species exhibiting free growth and of recurrently flushing species, which continue to expand their shoots late into the summer (Kozlowski and Clausen 1966).

Interpretation of the impact of environmental stress on tree growth is also complicated by environmental preconditioning. Rowe (1964) emphasized that, in the morphogenesis of a plant, early ecological influences can carry through to expression in later stages of development and behavior. Sensitive periods for preconditioning seem to be at the time of initiation and formation of buds and seeds as well as at the time when growth begins following a dormant or resting stage. Rowe concluded that it may be difficult to separate "heritable" from environmentally induced effects because of the probability of preconditioning before the seed was collected.

INTERNAL CONTROL OF GROWTH

Knowledge of how unstressed trees grow is essential to understanding how growth is reduced in stressed trees. Growth is the end result of cell division, cell expansion, differentiation, and morphogenesis. These sequential changes require a balanced supply of food, water, minerals, and hormonal growth regulators in meristematic regions. Regulation of growth involves close interdependency between roots and shoots. Roots depend on leaves for carbohydrates and hormonal growth regulators. Shoots, in turn, depend on roots for supplying water and mineral nutrients. In stressed trees dysfunctions occur in regulatory processes that control the availability of substances required to maintain growth at a maximum rate. Furthermore, growth inhibition or death of trees following insect attack or fungus invasion is preceded by abnormal physiological events.

The paramount importance of physiological processes in controlling tree growth can be illustrated by responses of trees to changes in photoperiod. We know that cessation of shoot growth and development of dormancy often have been linked to the shortening days of autumn. However, the short days do not directly promote shoot dormancy. We do know that leaves perceive the change from long to short days and produce a growth inhibitor that is transmitted from the leaf to the growing point. This inhibiting substance then plays an important role in inducing dormancy. This emphasizes that environmental stresses that do not cause direct injury to trees alter growth through a sequential physiological change pathway. Hence environmental changes and stresses are very remote from the final growth response.

Environmental stresses often set in motion a sequential and complicated series of metabolic disturbances, rather than a simple change in only one process such as photosynthesis, as is sometimes supposed. For example, cold soils or drought may decrease absorption of water and minerals. Decreased absorption of water then induces stomatal closure, which leads to reduced synthesis of carbohydrates and hormones, and their subsequent basipetal translocation to stem and root tissues. This sequence of events further decreases root growth, which decreases absorption of water even more, and so on. Hence the physiological impact of an environmental stress imposed on one part of a tree is transmitted to distant organs and tissues and eventually adversely affects the entire tree (Kozlowski 1969).

Carbohydrates

The importance of photosynthesis for tree growth cannot be overemphasized because more than two-thirds of the dry weight of trees consists of transformed sugars. Hence, growth of trees depends fundamentally on synthesis of carbohydrates and their immediate use or conversion to storage forms, followed eventually by reconversion to soluble forms, phloem loading, flow under pressure to sites of meristematic activity, and assimilation into new tissues. Carbohydrates are the chief constituents of cell walls. Large amounts of carbohydrates are oxidized in respiration to produce the energy needed for various metabolic processes and synthesis of new protoplasm.

Photosynthesis can be broken down into several component processes including: (1) trapping of light energy by chloroplasts; (2) splitting of water and release of high energy electrons and O_2; (3) electron transfer leading to generation of chemical energy in the form of ATP and reducing power as $NADPH_2$; and (4) final steps involving expenditure of energy of ATP and reducing power of $NADPH_2$ to fix CO_2 molecules in phosphoglyceric acid and reduce it to phosphoglyceraldehyde, and finally convert this compound into more complex carbohydrates, such as sucrose, starch, cellulose, and hemicellulose.

In recent years a great deal of attention has been given to two different specific biochemical pathways of photosynthesis. These will be mentioned briefly because of their important relation to plant responses to environmental stresses.

C_3 Plants. Most higher plants and almost all trees, except some mangroves and a few others, follow the Calvin cycle in which a 5-carbon sugar, 1,5 bisphosphate (RuBP) is carboxylated to form two molecules of 3-phosphoglycerate in a reaction catalyzed by RuBP carboxylase. At each turn of the Calvin cycle a molecule of CO_2 enters and is reduced while a molecule of RuBP is formed and six molecules of RuBP are reformed to continue the cycle.

C_4 Plants. In contrast to most woody plants, many crop plants of the tropics and subtropics, including corn, sorghum, and sugar cane follow the Hatch-Slack pathway of photosynthesis in which the 4-carbon acid, oxaloacetic acid, is formed when CO_2 is added to the 3-carbon compound, phosphoenolpyruvate (PEP). The reaction is catalyzed by PEP carboxylase. In C_4 plants, malic and aspartic acids are formed and broken down to produce CO_2 and pyruvic acid. The CO_2 is transferred to RuBP of the Calvin cycle and pyruvic acid reacts with ATP to form more PEP. Hence photosynthesis of C_4 plants differs from that of C_3 plants only in the initial steps.

The chloroplasts of C_3 plants are similar throughout the palisade parenchyma, and the Calvin cycle operates in each cell. In contrast, in C_4 plants only certain specialized leaf cells fix most of the CO_2. Leaves of C_4 plants have vascular bundles surrounded by large bundle sheath cells containing prominent chloroplasts and starch grains. However, starch usually is absent in chloroplasts of mesophyll cells. Atmospheric CO_2 is fixed only in the mesophyll cells by the PEP system and these cells lack the Calvin cycle. Malic and aspartic acids that form in the mesophyll cells are translocated to the bundle sheath cells where the Calvin cycle operates and CO_2 is added.

Unfortunately, virtually all woody plants, with their C_3 photosynthesis, are less well adapted than C_4 plants to undergo stress conditions of drought and higher temperatures. The optimum temperature and light intensity are higher for C_4 than for C_3 plants. Another disadvantage for C_3 plants is that they use atmospheric CO_2 less efficiently than C_4 plants do. In C_3 plants the action of RuBP carboxylase is inhibited because O_2 competes with CO_2 causing a light stimulated respiration, called photorespiration. C_4 plants do not exhibit photorespiration. In C_4 plants two separate carboxylation reactions lead to efficient CO_2 absorption even at low CO_2 concentrations followed by transfer to the Calvin cycle which produces the photosynthetic products needed in growth.

The loss of as much as a half of the photosynthate in photorespiration appears to be important in limiting growth of trees under environmental stress. However, in favorable environments there is a question as to whether growth is actually limited by the rate of photosynthesis per unit of leaf area.

The relation of photosynthetic capacity to growth is complicated. Photosynthetic rates vary appreciably among species and particularly high rates have been confirmed for such fast growing trees as *Eucalyptus,*

Populus, and *Pseudotsuga.* Photosynthetic capacity also varies greatly among varieties, clones, and provenances of the same species.

Much interest has been shown in using photosynthetic capacity as an index of growth potential of trees. However, both high and low and even negative correlations between photosynthetic rates and growth of trees have been shown. Low correlations and negative correlations between photosynthesis and growth potential often are due to inadequate sampling of photosynthesis, usually determined as CO_2 absorption over a short period of time. In addition to photosynthetic rate at a particular time, growth potential of trees depends on the seasonal pattern of photosynthesis and duration of growth, the relation of photosynthesis to respiration, and the partitioning of photosynthate within the tree.

A tree with a high rate of photosynthesis at one stage of its seasonal cycle may have a low rate at another stage. From April to August a highland ecotype of *Pinus silvestris* had a higher rate of photosynthesis than a lowland ecotype. Thereafter the rate of the highland ecotype decreased rapidly and was lower during the autumn than the rate of the lowland ecotype (Zelawski and Goral 1966). Similarly, the rate of photosynthesis of several *Pinus banksiana* provenances changed with time. One provenance with a very high rate in July had one of the lowest rates in November. Provenances with very high rates in October and November also had the highest growth rates (Logan 1971). Such observations emphasize that prediction of dry matter increment from measurements of photosynthesis should be based on rates and rate-duration aspects of photosynthesis.

The relation between photosynthetic capacity and wood production sometimes is also complicated by partitioning of photosynthetic products. The relation may be high during a year of vegetative growth and low during one of reproductive growth when much of the photosynthate is channeled to fruit and seed production. Many investigators have shown that the amount of fruit or seed crop is negatively correlated with vegetative growth (see Kozlowski 1971b).

Water

Both the distribution and growth of forest trees depend on a favorable internal water balance. Water is essential as a constituent of physiologically active cells, a reagent in photosynthesis and hydrolytic processes, and as a solvent in which solutes move from cell to cell. An essential role of water is maintaining turgor of guard cells and photosynthetically active cells. The importance of high turgor to photosynthesis cannot be overemphasized because the rate generally declines when leaves are only slightly dehydrated. When trees are subjected to a drought and then rewatered, the rate of photosynthesis often fails to return to normal for a very long time because of damage to stomata and chloroplasts. Eventually even mild water deficits inhibit tree growth. Furthermore the loss of growth frequently attributed to competition or root injury is often traceable to decreased absorption of water leading to desiccation of the tree crown. The effect of water balance on growth of trees is discussed in detail in the next lecture.

Minerals

Mineral elements have many functions in forest trees. They are constituents of plant tissues, coenzymes, osmotic regulators, catalysts in biochemical processes, constituents of buffer systems, and regulators of membrane permeability.

A mineral element is classified as essential if a plant cannot complete its life cycle without it and if it is a component of a molecule of an essential plant constituent. Elements considered essential for trees include nitrogen, phosphorus, potassium, calcium, sulfur, magnesium, iron, copper, zinc, boron, molybdenum, and chlorine. The first six of these, the major elements, are required in relatively large amounts; the last seven, the minor elements, are required in very small amounts.

Mineral elements stored in a seed, along with carbohydrates and fats, provide an emerging seedling with its initial supply of nutrients, but for most temperate zone species these stored minerals can sustain the plant for only a few days or weeks. Once the mineral nutrients of a seed are exhausted, additional minerals must be supplied from sources in the soil. Dissolved mineral nutrients in precipitation or applied as liquid fertilizers are also absorbed by leaves.

Mineral elements represent a very small proportion

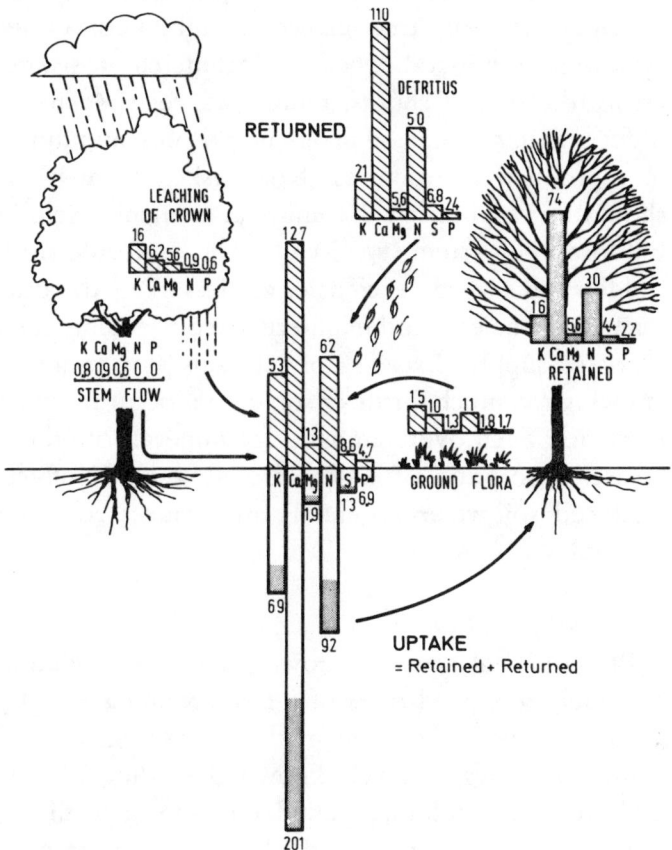

Fig. 3. Cycling of minerals in a mixed hardwood forest. Amounts are given in kg per hectare per year. Yearly uptake of minerals by the stand is shown by amounts retained by plants (dotted bars) and those returned to the soil (cross hatched bars) by leaching, stem flow, and litter. Reprinted, by permission, from Larcher 1975, fig. 80.

of tree biomass but without an adequate supply of any essential element, tree growth is inhibited. Other symptoms of mineral deficiency include chlorosis, necrosis, dieback of shoots, gummosis, retarded tissue differentiation, decrease in vascular tissues, and injury to reproductive structures (Kozlowski 1971a, b). It should be emphasized that mineral elements cycle in the forest ecosystem (fig. 3) and that, on fertile sites, sudden and severe deficiencies are less likely to occur than deficiencies of other internal requirements such as water supply. Excesses of minerals as a result of applying too much fertilizer sometimes occur in forest nurseries. Such overabundance of minerals often decreases absorption of water through osmotic effects (reduced soil-water potential) and injures roots by plasmolysis.

Hormonal Growth Regulators

Physiological activity, growth, and differentiation of widely separated parts of a tree are integrated by a complex hormonal system. Five broad groups of plant hormones are well known, including auxins, gibberellins, cytokinins, ethylene, and growth inhibitors such as abscisic acid (ABA). Although each of these hormones has been associated with specific plant processes, each variously influences cell division, increase in cell size, and differentiation. Much early research ascribed changes in growth to effects of individual hormones. However, the current view is that growth control is regulated by synergistic effects of two or more hormonal growth regulators.

The effects of environmental stresses often are mediated, at least in part, through alterations in hormone balances. For example, short days may induce bud dormancy by changing the hormone balance in favor of growth inhibitors over growth promoters (Kozlowski 1971a). Both drought and flooding stimulate production of ethylene which induces leaf epinasty and abscission. A nutrient deficiency sometimes checks growth by inhibiting hormone synthesis. For example, reduced growth of some species under low nitrogen availability sometimes is due largely to hormone deficiency rather than to lack of nitrogen per se (Wareing 1974).

Important Environmental Stresses

Many environmental factors affect growth of trees. Among the most important are light, temperature, water supply, mineral supply, and air pollutants. These will be discussed separately. Water supply will be treated in considerable detail in the next essay.

Light Stress

It is well known that tree growth is influenced by light intensity, light quality, and periodicity, with the effects of light intensity probably most important. Light intensity is an important factor in competition and has much to do with stratification of trees into crown classes and eventual phasing out of very suppressed trees. Over the years we have learned that species vary widely on the basis of their shade tolerance and capacity for natural pruning of lower branches. Overall, however, we have been very complacent about the impact of light stress on maximum productivity. Only in very recent years has serious thought been given to investigating and characterizing the central features that account for shade tolerance. Such information might lead to increased productivity through modification by tree breeders of shade tolerance of desirable species and genetic materials.

Light transmission through trees decreases rapidly as the density of tree crowns increases up to about 35 percent, and then transmission decreases more slowly. Light intensities under open even-aged pine stands of the temperate zone generally approximate 10 to 15 percent of full light. Light intensities under temperate zone hardwoods are even lower, commonly 1 to 5 percent of that in the open. Light intensities under tropical forests are still lower, often less than 0.5 percent of that in the open (Spurr 1964). Hence, deep shading as a stress factor is always with us.

It is important but not easy to evaluate the physiological implications of such quantitative comparisons among species. This is because sunflecks temporarily increase relative illumination to very high values. For example, in some Nigerian forests sunflecks occupy about one-fourth of the surface of the forest floor and they account for four-fifths of the incident radiation at midday (Evans 1956). Sunflecks appear to play an important role in temporarily increasing photosynthesis and increasing competitive capacity of some shade-adapted species (Leopold and Kriedemann 1975).

Open-grown trees are too often considered not to be undergoing light stress. Yet any tree, shade tolerant or not, undergoes a great deal of mutual shading because of the stacked arrangement of its leaves. The shape of the crown and density of foliage influence the radiation profile from the surface of the crown to the interior (fig. 4). In the interior of the crown of some species the light intensity may be less than 1 percent of that in the crown periphery. The interior of dense tree crowns often lacks foliage.

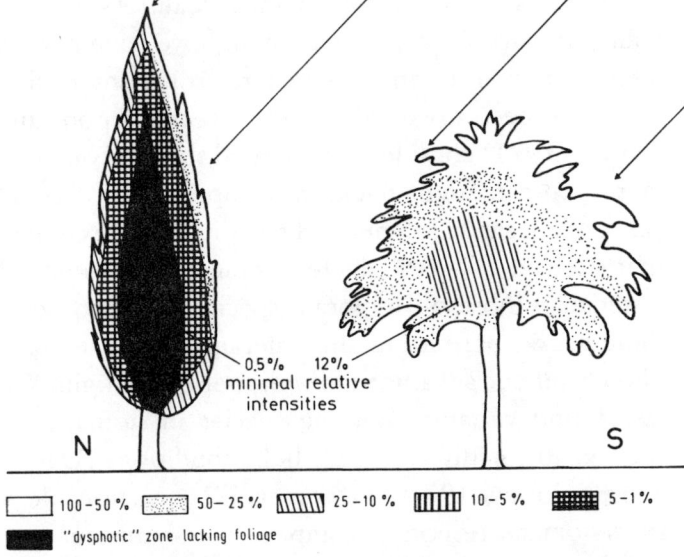

Fig. 4. Variations in light intensity in different parts of the dense crown of a cypress (left) and the open crown of an olive tree (right) at noon on a summer day. Light intensity is given as a percent of that in the open air. Reprinted, by permission, from Larcher 1975, fig. 7.

Although shade tolerance varies widely among species, it also varies with age of trees and environmental conditions. Trees tend to show the greatest shade tolerance in their youth, and those on good sites and in the southern part of their range are more shade tolerant than those on poor sites or in the northern part of their range.

If geneticists and tree breeders are to increase the capacity for shade tolerance of given plant material it would seem imperative as a first step to identify some central physiological or biochemical feature(s) that accounts for shade tolerance. Although we have not yet identified such a common feature, some interesting data have been obtained in recent years. Woods and Turner (1971) considered that rapid stomatal opening in response to light flecks was important in overall conferral of shade tolerance. They found, for example, that stomata of a few shade tolerant species opened faster than those of intolerant species, in response to light flecks, permitting the tolerant species to take photosynthetic advantage of short periods of light. We also found variations among species in stomatal responses to rapidly changing light conditions (Davies and Kozlowski 1974). However, it is questionable if such stomatal responses comprise the significant central feature in the shade tolerance mechanism. In a number of species poor correlations exist between shade tolerance and stomatal responses to rapid changes in light intensity (Pereira and Kozlowski

1977). It should be remembered that shade tolerance ratings for different species are relatively stable whereas stomatal responses for these same species are modified by light- and water-stress preconditioning as well as temperature. Furthermore, both stomatal opening and closing are much slower in chlorotic plants than in green ones, emphasizing the importance of nutritional status (Davies and Kozlowski 1974).

Durzan (1971) emphasized the importance of a metabolic base in the shade tolerance mechanism. He showed, for example, that much greater changes occurred in amounts and numbers of amino acids in response to shading the intolerant *Pinus banksiana* over the shade tolerant *Picea glauca*. Some very interesting research by Alberte, McClure, and Thornber (1976) showed that shade-grown leaves contain a higher proportion of chlorophyll a/b protein than sun-grown leaves. This difference accounts for greater efficiency of shade leaves in harvesting light, but it also results in their light saturation at a lower light intensity. More study is now indicated on the nature of light-harvesting pigment in shade tolerant and intolerant forest trees.

Temperature Stress

Most plant processes and growth attain a maximum in the range of 20 to 35°C depending on species. Higher or lower temperatures inhibit growth.

Low Temperature Stress. Spring and autumn frosts are responsible for several forms of freezing injury to insufficiently hardened trees. Both the nature and amount of injury vary with the duration of a freeze and the degree of cold hardiness in plant tissues at the time of the freeze. A common response is killing of shoots, especially those of late-season growth flushes such as lammas shoots, or as a result of irrigation or application of fertilizers late in the growing season, thereby prolonging shoot growth. Injury to the cambium by freezing may produce "frost rings" characterized by abnormally developed cambial derivatives and displaced wood rays. Another effect is the development of stem lesions and cankers, especially when alternate freezing and thawing occur during the winter and early spring. Death of small, physiologically important roots and frost heaving of tree seedlings in nurseries are other manifestations of low temperature injury.

Freezing Injury. Trees may be injured by intracellular freezing of cell contents or dehydration of cells as a result of extracellular freezing. Plants usually are killed when ice crystals form within cells, but ice crystals in spaces between cells do not always cause death. Slow cooling of cold hardened plants induces ice formation in intercellular spaces, movement of water from cells to the ice nuclei, increase in concentration of cell sap, and lowering of its freezing point. Rapid cooling of cold hardened plants may cause injury when cells are not dehydrated rapidly enough and intracellular freezing occurs.

Cold Hardiness. Many temperate zone trees can survive very low winter temperatures without injury, but the same trees are killed if they are artificially exposed to temperatures only a few degrees below 0°C in the summer. Cold hardiness varies greatly among forest trees. Plants native to warm regions often cannot be moved to cold regions because they do not develop sufficient hardiness to cold. Northern species, including *Populus tremuloides, P. balsamifera, Betula papyrifera,* and *Larix laricina* resisted freezing in midwinter to −80°C or lower. Most Northern Rocky and Western Mountain conifers, including *Pinus ponderosa, P. monticola, P. contorta, P. jeffreyi, Picea pungens, P. engelmannii, Abies lasiocarpa, A. concolor,* and *Larix* were cold hardy between −60° and −80°C. In contrast, southern pines, *Quercus virginiana* and *Magnolia grandiflora* developed winter hardiness down to only −15°C (table 1).

Cold hardiness also varies greatly within a species, with southern ecotypes and climatic races not able to develop enough cold hardiness when moved northward. For example, Squillace and Silen (1962) found that northern provenances of *Pinus ponderosa* were much more resistant than southern ones to freezing. Better development of cold hardiness of northern provenances often is associated with early setting of buds. Campbell and Sorenson (1973) found that southern seed sources of *Pseudotsuga menziesii* set buds later and were less cold hardy than northern sources when both were grown at Corvallis, Oregon. Southern provenances were also less cold hardy than northern ones, even when they set buds in the same week. Such marked variations in cold hardiness of different provenances apparently do not occur in all species. For example, Smithberg and Weiser (1968) found that

TABLE 1

VARIATIONS IN FREEZING RESISTANCE OF NORTH AMERICAN TREE SPECIES AND MINIMUM TEMPERATURES AT NORTHERN LIMITS OF NATURAL RANGES OR ARTIFICIAL PLANTINGS

Relative Hardiness Classification	Representative Species	Average Minimum Temperatures at Northern Limits of Growth (°C)		Observed Freezing Resistance (°C)
		Natural Range	Artificial Plantings	
Tender evergreen species	*Quercus virginiana*	−3.9 to −6.7	−9 to −12	−7 to −8
Hardy evergreen species	*Magnolia grandiflora*	−9 to −12	−18 to −20	−15 to −20
Hardy deciduous species	*Liquidambar styraciflua*	−18 to −20	−26 to −29	−25 to −30
Very hardy deciduous species	*Ulmus americana*	−37 to −46	−40 to −43	−40 to −50
Extremely hardy deciduous species	*Betula papyrifera*	below −46	below −46	below −80
	Populus deltoides	−32 to −34	−37 to −45	below −80
	Salix nigra	−32 to −34	−37 to −45	below −80

SOURCE: Reprinted, by permission, from Sakai and Weiser 1973, table 11. © 1973 by the Ecological Society of America.

northern clones of *Cornus stolonifera* hardened somewhat earlier than southern clones, but all of them developed the capacity to resist temperatures of −90°C by early December.

Cold hardiness also varies appreciably among different organs and tissues. Fortunately roots are well protected by soil and by snow cover because they are extremely sensitive to cold. Opening vegetative buds, flower buds, and young conelets also are highly susceptible to frost injury. In stems the xylem parenchyma cells of sapwood develop less cold hardiness than cells of the cambium, phloem, and cortex.

Changes During Development of Cold Hardiness. Cold hardiness in trees develops in three successive stages. The first stage, which is set in motion before the first frost of the season, involves cessation of growth and several metabolic changes that condition the tree to respond to low temperatures during the second stage of acclimation to cold. The first stage appears to be induced by decreasing day length and the second stage by the first frost. The third stage is triggered by very low temperatures. Some trees that are well into the third stage can withstand temperatures as low as −200 °C, but if thawed for only a few hours they lose their capacity to survive such very low temperatures.

There is some disagreement about whether or not dormancy is a prerequisite for development of cold hardiness. Weiser (1970) believes that growth cessation rather than dormancy is the major factor in initiation of cold hardiness. Glerum (1976), however, emphasized that both development and loss of cold hardiness lag behind induction and loss of dormancy. His view is that dormancy is a prerequisite for cold hardiness, at least in some species.

Several internal changes occur during acclimation of tissues to cold and some of these changes, alone or together, appear to account for cold hardiness. In most trees sugars build up and may increase cold resistance by accumulating in vacuoles, decreasing the amount of ice formed, decreasing cell dehydration associated with freezing, and by some not-well-understood metabolic "protective changes" (Levitt 1972). Another important change during development of hardiness is an increase in water soluble proteins, especially during late stages of acclimation. Changes in soluble proteins may involve alterations in water binding proteins and a decrease in free water in cells, making formation of intracellular ice less likely. Other changes include increases in RNA, reduction of tissue hydration, increase in lipids, and increase in membrane permeability (Alden and Hermann 1971).

Chilling Injury. Many chilling-sensitive trees of the tropics undergo injury at low temperatures a few degrees above freezing, usually in the range of 0–5°C but sometimes up to 10°C. Such injury often is not evident until after days or weeks of exposure to the chilling temperature. Chilling injury, which varies greatly among species and cultivars, may be expressed in lesions, susceptibility to decay organisms, growth inhibition, and death of trees. The exact mechanisms of chilling injury are not fully understood, but imbalances in photosynthesis, respiration and metabolism, as well as accumulation of toxins and membrane effects appear to be involved. A common response is solidification of membrane lipids of chilling-sensitive species, followed by development of cracks or channels in cell membranes leading to leakage of ions and death of cells (Lyons 1973).

High Temperature Stress. Unusually high temperatures often injure trees directly and indirectly. Temperatures in the 45–60°C range generally cause direct injury; somewhat lower but still high temperatures usually induce indirect injury. Direct injury occurs rapidly after only brief exposure to heat, whereas indirect injury may not be evident for many days after trees are exposed to the high temperatures.

Direct heat injury often appears as stem lesions in seedlings and mature trees. Bark desiccation (sunscald, bark scorch) often is a response of smooth-barked transplanted trees or those exposed by thinning. Such responses are accentuated more by large diurnal temperature fluctuations than by high temperature alone.

Indirect high temperature injury may involve metabolic changes such as loss of carbohydrate reserves or formation of toxic products. High temperatures often result in depletion of food reserves because photosynthesis declines rapidly after a critical high temperature is reached but respiration continues to increase above this temperature. Another dysfunction associated with high temperature is breakdown of protoplasmic proteins. As mentioned earlier, high temperature may also injure and kill trees by desiccation resulting from accelerated transpiration. This will be discussed further in the next lecture.

Mineral Stress

Until recent years, the loss of soil fertility as a result of harvesting methods has not attracted a great deal of attention. Both long rotations and removal of only parts of trees have depleted only small amounts of the nutrient capital from many temperate zone sites. Nutrient losses range from about one to three pounds per acre for phosphorus, ten pounds for nitrogen and potassium, and even more for calcium in some hardwood stands. Such removals are more than replaced by natural weathering of soil minerals together with low levels of nitrogen fixation and atmospheric contributions. These views are in accord with the fact that many European forests have been harvested regularly for 500 years. However, the recent surge of interest in complete tree harvesting poses questions about more serious mineral losses on various sites, especially those deficient in minerals even before complete tree harvesting was practiced on them.

From an economic standpoint complete tree harvesting is very attractive. Spurr (1976) estimated that yield can be increased by about 6 percent if all the wood in the main stem is used, by 15 percent if all the bark in the stem is used, and by 42 percent if all branches are harvested and used (an increase of 54 percent over the merchantable wood in the stem). If the root system is used as well, the yield is potentially about twice that of the merchantable stemwood only.

Complete tree harvesting plus use of short rotations, a combination called fiber farming, may not be advisable in the long term, on certain sites at least, except with a fertilizer program. In some areas where addition of fertilizer may not be practical, considerable attention to nutrient management will be mandatory.

The impact of fiber farming will vary for different sites. The extent to which loss of mineral nutrients is increased by this practice will depend on the distribution of biomass components, which varies with tree species, tree age, and site. Leaves usually have the highest concentration of minerals followed in order by small roots and twigs, branches and large roots, and stems. Hence, the percentage increase in loss of mineral nutrients as a result of conversion to complete tree harvesting will be less for species with small crowns and small leaf biomass than for those with large crowns. It also will be less for old stands than young ones (short rotations). With increasing tree age the proportion of minerals increases in the stem and decreases in the crown. On moist and fertile sites having trees with large leaf and branch biomass the increase in nutrient loss will be greater than for the same species (with small crowns) growing on nutrient- and water-deficient sites. The loss of mineral nutrients will also be influenced by the concentration of minerals in harvested tissues. It has been estimated that converting from one thirty-year rotation for *Populus tremuloides* to three ten-year rotations, all with complete tree harvesting, will increase depletion of N, P, K, and Ca by 345, 239, 234, and 173 percent, respectively (Boyle

1975). Some examples of the effect of forest type and age of trees on losses of mineral nutrients that may be expected in conversion from conventional to complete tree harvesting are shown in Tables 2 and 3.

TABLE 2

PERCENTAGE INCREASE IN DEPLETION OF PLANT NUTRIENTS IN HARVESTED MATERIALS ACCOMPANYING THE CHANGE FROM CONVENTIONAL TO WHOLE-TREE HARVESTING

Forest Type	% Increase				
	Above Ground Biomass	N	P	K	Ca
Hemlock–cedar < 500 years old	43	165	117	77	95
Pine, 125 years old	15	53	54	14	15
Spruce–fir < 350 years old	25	116	163	32	50
Hemlock–fir < 550 years old	20	86	67	48	48
Spruce, 65 years old	99	288	367	236	179
Cottonwood, 9 years old	—	116	100	74	68

SOURCE: Reprinted, by permission of the author, from Kimmins 1977, table 1.

Tree growth often is influenced more by the rate of mineral cycling than by the actual mineral nutrient capital of the site. If nutrients cycle rapidly between trees and soil, productivity may be high even when the nutrient capital is modest. The rate of mineral turnover depends on the rate of decomposition of organic matter. If complete tree harvesting induces rapid decomposition and mineral turnover, the losses in productivity may be balanced by accelerated circulation of residual minerals. By comparison, most of the minerals on sites characterized by slow decomposition may be in the trees themselves, and nutrient cycling may be restricted to mineral elements in litterfall and leachates from leaves. Under such conditions harvesting of tree crowns can remove a large fraction of the nutrient pool and adversely affect growth of the subsequent tree crop.

TABLE 3

EFFECT OF AGE OF STAND ON PERCENTAGE INCREASE IN NUTRIENT LOSSES ON CONVERSION FROM CONVENTIONAL TO WHOLE-TREE HARVESTING

		% Increase			
Species	Age (Years)	N	P	K	Ca
Spruce	18	195	233	161	206
	50	114	115	26	40
	85	91	104	42	29
Pine	39	164	200	140	88
	44	124	133	108	84
	75	77	67	56	59
Pines, average	50	—	156	104	100
	100	—	87	59	52
Other conifers, average	50	—	170	127	138
	100	—	87	56	59
Hardwoods, average	50	—	122	92	67
	100	—	69	47	37

SOURCE: Reprinted, by permission of the author, from Kimmins 1977, table 2.

Shrubs and herbaceous plants usually have higher mineral concentrations than trees, and their litter usually decomposes faster than that of overstory trees. It follows that complete tree harvesting in forests with well-developed shrub layers will remove less of the cycling minerals than in dense conifer plantations that essentially lack a shrub layer.

The exact impact of complete tree harvesting will vary with mineral requirements of the subsequent crop. For example, removal of nutrients may not be serious, at least for several rotations, if the subsequent crop will consist of pines that have relatively low mineral requirements.

Still another factor influencing the impact of whole tree harvesting is the magnitude of other losses of mineral nutrients. For example, burning of slash can result in appreciable losses of nitrogen, sulfur, and potassium. Hence, under certain conditions losses of minerals under complete tree harvesting may not be as great as in conventional logging with slash burning.

The impact of fiber farming will vary with the frequency of loss of mineral nutrients. An appreciable sudden depletion of minerals will only temporarily reduce tree growth if natural processes replace minerals and the next harvest is delayed. In contrast, a serious, progressive loss of nutrients can occur from a small loss if repeated too frequently. The shorter the rotation, the greater is the risk of mineral depletion. Also the greater the amount of material removed, the greater is the risk of mineral depletion for a given rotation and rate of replacement. Hence, long rotations may be required for complete tree harvesting on infertile sites. If, however, the rate of mineral replacement is naturally high or maintained with fertilizers, short rotations may not degrade the site. More research is needed on the impact of whole-tree harvesting and short rotations on mineral relations of forests.

Pollution Stress

One of our most immediate problems is how to cope with the impact of environmental pollution. In relatively recent times the total amount and complexity of toxic pollutants in the environment have increased dramatically. Prior to the recent energy crisis it was predicted that by the year 2000, total emission of gaseous pollutants would be increased by about three-fourths (Houston 1970). The increased use of coal accompanying the energy crisis will increase sulfur pollutants in the air. Hence we can no longer expect trees to grow in clean air.

Before 1940 the major air pollutants affecting trees were smoke and sulfur dioxide. Now, however, trees are adversely influenced by a wide array of pollutants, including gases, particulates, agricultural chemicals, and radioactive materials. In the United States sulfur dioxide, ozone, hydrogen fluoride, and peroxyacetyl nitrate are considered the most important air pollutants. Sulfur dioxide and ozone probably cause more injury to trees than all other air pollutants combined. Fluorides often cause extensive damage around large point sources such as smelters and fertilizer plants but, on balance, fluorides cause less damage than sulfur dioxide and ozone (Mudd and Kozlowski, 1975).

The tragedy of the pollution problem is that it is man-made and the result of years of apathy. The motor vehicle is the most important source of air pollution and industrial sources are a distant second, emitting about a fourth as much as transportation does. Generation of steam and electric power produces slightly less pollution than industry. Space heating emits only about 10 percent as much pollution as transportation. The composition of pollutants from various sources differs greatly, industry emitting the most diversified pollutants.

Sources of Air Pollutants. Air pollutants can be divided into primary and secondary types. Primary pollutants, such as SO_2 and HF, originate at the source in a form that is toxic to plants. Secondary pollutants, such as peroxyacetyl nitrate (PAN) and ozone, develop as a result of reactions between pollutants that originate from a source.

Sulfur dioxide is emitted principally from combustion of coal. It is also produced during production, refining, and use of petroleum and natural gas; manufacture and use of sulfuric acid and sulfur; and smelting and refining of ores. Fluorine compounds originate from aluminum reduction processes, manufacture of phosphate fertilizer, steel manufacturing plants, brick plants, and refineries.

The most important sources of ozone appear to be photochemical reactions in polluted atmospheres. Nitrogen oxides emitted into the atmosphere by automobiles, industries, and utilities may react in the presence of sunlight to form ozone.

Peroxyacetyl nitrates are secondary pollutants

formed by photochemical reactions involving the primary pollutants emitted in automobile exhaust.

The most important oxides of nitrogen in the atmosphere are nitric oxide (NO) and nitrogen dioxide (NO_2). During combustion some of the nitrogen in the air is oxidized to NO and a rather small amount to NO_2. Important particulates include cement-kiln dust, flourides, lead particles, soot, magnesium oxide, iron oxide, foundry dusts, and aerosols. For more details on sources of pollutants the reader is referred to Mudd and Kozlowski (1975).

Effects of Air Pollution. Air pollution is responsible for tremendous losses of tree growth and changes in the structure and function of forest ecosystems. Trees are eliminated first by low dosages of pollutants. As duration of exposure increases, tall shrubs are eliminated, followed in order by lower shrubs, herbs, and mosses and lichens (Woodwell 1970).

Extensive damage to forest trees by pollution from point sources has been reviewed by Miller and McBride (1975). They describe in detail the extent of injury to various species of forest trees caused by variously located smelters, mills, and power generating plants. The use of tall stacks in recent years has reduced acute air pollution injury, but chronic exposure of forests to air pollution is expected to continue. Weinstein (1975) believes that chronic responses may be ecologically and economically more important than acute responses.

Growth of trees is adversely affected by SO_2 at levels below those that cause visible symptoms of injury. Phillips, Skelley, and Burkhart (1977a,b) showed that growth of *Pinus taeda* and *P. strobus* was reduced by low levels of SO_2. Houston and Dochinger (1977) reported reduction in seed production by *Pinus resinosa* and *P. strobus* by SO_2 levels that were too low to cause visible injury.

In addition to killing plants, atmospheric pollutants adversely affect plants in various ways. Pollution injury is most commonly classed as acute, chronic (chlorotic), or hidden. In acute injury collapsed marginal or intercostal leaf areas are noted which at first have a water-soaked appearance. Later these dry and bleach to an ivory color in most species, but in some they become brown or brownish red. These lesions are caused by sudden absorption of enough gas to kill the tissue. Chronic injury involves leaf yellowing which may progress slowly through stages of bleaching until most of the chlorophyll and carotenoids are destroyed and interveinal portions of the leaf are nearly white. Chronic injury is caused by sudden absorption of an amount of gas somewhat insufficient to cause acute injury, or it may be caused by absorption over a long period of time of sublethal amounts of gas. Sometimes the effects of different pollutants are separable and at other times indistinguishable. Solberg and Adams (1956) found histological responses of dicotyledonous plants to hydrogen fluoride and sulfur dioxide to be indistinguishable. Costonis (1970) did not find any histological differences of diagnostic significance between SO_2 and O_3 injury to *Pinus strobus*.

The histological changes most commonly noted in pollution-injured leaves include plasmolysis, granulation or disorganization of cell contents, cell collapse or disintegration, and pigmentation of affected tissue. Commonly physiological activity of affected plants is impaired before any external symptoms are visible. For this reason many investigators referred to "hidden," "invisible," or "physiological" injury of pollutants. Many years ago three criteria were established for such invisible injury: (1) it involved a disturbance of the life of the plant that was eventually expressed as an effect on growth, (2) the disturbance was not evident to the naked eye, and (3) it was present where plants underwent prolonged exposure to concentrations of pollutants that did not produce visible markings. Thomas (1961) redefined "invisible" injury as the reduction of photosynthesis below the level expected from the amount of leaf destruction. Invisible injury was considered to have two main characteristics: (1) it did not occur unless a threshold concentration of a pollutant was exceeded, and (2) it had a certain magnitude and a limited duration once the exposure to the pollutant ceased. These characteristics were determined by plant species and variety.

A complication in evaluating the physiological mechanism of pollution injury is that such factors as light, water, temperature, and mineral nutrition affect the response of plants to pollutants. Still another complication is that in the field more than one pollutant often is responsible for injury. For example, both ozone and SO_2 are responsible for the physiogenic needle disorder of *Pinus strobus* known as "chlorotic dwarf." Even tolerable levels of a single pollutant may injure plants when present with another pollutant at an equally low level.

Air pollutants appear to be relatively nonspecific agents that have many sites of action (Mudd and Kozlowski 1975). They affect many metabolic processes. The effect of a pollutant depends on its concentration in a cell as well as on metabolic patterns in cells. Most air pollutants decrease photosynthesis directly, or indirectly by causing loss of photosynthetic tissues (through leaf abscission, chlorosis, necrosis) and by affecting stomatal aperture. The literature on specific effects of SO_2 on metabolism is growing rapidly. SO_2 affects many enzymes of photosynthesis, causes breakdown of chlorophyll, and affects membranes. SO_2 enters leaves through the stomata. It reacts with the wet surfaces of mesophyll cells to form the very toxic sulfite which, in high concentration, rapidly kills cells. At low levels sulfite is converted to sulfate which is much less toxic than sulfite. Ozone appears to cause damage in at least three ways: (1) by interfering with mitochondrial activity, (2) by affecting membrane permeability, and (3) by inhibiting photosynthesis. Fluoride is absorbed from the air and translocated to leaf tips and margins where its high concentration rapidly kills cells. The effects of PAN may be similar to those of ozone.

Acid Rain. It has been known for a long time that many harmful effects of air pollution are exerted on forest trees located relatively close to the site of emission. In relatively recent times we have also become aware that some atmospheric pollutants and their reaction products are dispersed by meteorological processes and adversely affect trees very far from the source of emission.

Rain formed in an atmosphere relatively free of pollutants would be expected to have a pH of about 5.8. Both European and American studies show, however, that rains often are very acid. For example, the average pH of rainwater at the Hubbard Brook Experimental Forest in New Hampshire varied from 4.03 to 4.09 during 1965–71. Some rains with pH values as low as 2.1 have been reported in the northeastern United States, often many hundreds of kilometers from major sources of air pollution (Likens and Bormann 1974).

Very little research has been done in the United States on deposition of pollutants far from the source of emission and the biological consequences of acid rains. By comparison, extensive measurements have been made in Europe since the late 1940s. Such studies emphasize that some countries which do an excellent job of controlling pollution are victimized by pollution emanating in other countries. For example, more than three-fourths of the sulfur in rain deposited in Norway and Sweden is reported to originate largely in the industrialized regions of England and Central Europe. The sulfur compounds in rain are dispersed largely by wind (Braekke 1976).

Because of the European experience we need to be concerned with the fact that the pH of rain in much of the eastern United States was lower than 6 in 1956 and that the zone of highest acidity occurred in states with high sulfur emission (parts of Ohio, Pennsylvania, West Virginia, New York, and the New England states). By 1972–73 the area with an average pH of rain below 4.5 had extended to include parts of Mississippi, Alabama, Georgia, South Carolina, North Carolina, Kentucky, Virginia, and north into New England and Canada.

We do not have much good data on the effects of acid rains on forests in the United States. However, measurements in southern Sweden were made from 1896 to 1965 and indicated that growth of forest trees was reduced by 2 to 7 percent (average of 4 percent) between 1950 and 1965. Jonsson and Sundberg (1972) attributed these growth decreases to acidification.

European studies further show that: (1) crowns of forest trees filter out sulfur from rain, (2) bark acidity and developmental cuticular characteristics such as stomatal size and frequency are indicators of sulfur pollution, (3) plants on recently formed sandy soils or glacial outwashes are harmed more by acid rain than are plants on older, well-buffered soils with higher clay content and associated high base exchange capacity, (4) acid rains directly induce changes in leaf physiology, (5) after reaching the soil, acidic substances can cause alterations in root physiology and availability of essential cations, and (6) ammonia combines with sulfate ions in the presence of water in the atmosphere and hence tends to neutralize atmospheric acidity.

However, when the ammonia is absorbed by plants, both the ammonia and released sulfate ions contribute to the total acidification of forest ecosystems.

A number of studies show that simulated rain acidified with sulfuric acid inhibits seed germination, injures leaves, accelerates leaching of nutrients from leaves, increases erosion of leaf waxes, decreases mycorrhizal uptake of nitrogen, inhibits nodulation and nitrogen fixation by legumes, and decreases growth of trees. In forests most of the acid rain impinges on the leaves where ion exchange occurs. Hence the rain that reaches the soil is much less acid than that falling on leaves. In addition to being alert to direct effects of acid rain we need to be concerned about its potentially indirect effects such as leaching of nutrients from leaves and providing access to pathogens, insects, and biocides (Dochinger and Seliga 1975).

At present we do not begin to understand all the biological consequences and ramifications of acid rain on forest ecosystems of the United States. We know, however, from the European experience that we should devote more energy to measurement of long-term changes in the chemistry of air in North America and assessment of the biological and economic impact of acidified rain waters.

Pollution Resistance. There are wide differences among species and cultivars in resistance to air pollution (Tables 4 and 5). Although no plant is totally immune to pollution, enough genetic variation has been found in response to pollution to select for resistance. Success in developing high-yielding, genetically improved trees will depend on elucidating the specific nature of pollution resistance of various plant materials.

TABLE 4
RELATIVE SUSCEPTIBILITY OF TREES TO SULFUR DIOXIDE

Sensitive	Intermediate	Tolerant
Betula alleghaniensis	*Abies balsamea*	*Abies amabilis*
Betula papyrifera	*Abies grandis*	*Abies concolor*
Betula populifolia	*Acer negundo*	*Acer platanoides*
Fraxinus pennsylvanica	*Acer rubrum*	*Acer saccharinum*
Larix occidentalis	*Picea engelmannii*	*Acer saccharum*
Pinus banksiana	*Picea glauca*	*Juniperus occidentalis*
Pinus resinosa	*Pinus contorta*	*Picea pungens*
Pinus strobus	*Pinus monticola*	*Pinus edulis*
Populus grandidentata	*Pinus nigra*	*Pinus flexilis*
Populus tremuloides	*Pinus ponderosa*	*Quercus gambelii*
Salix nigra	*Populus balsamifera*	*Quercus palustris*
	Populus deltoides	*Quercus rubra*
	Populus trichocarpa	*Thuja occidentalis*
	Pseudotsuga menziesii	*Thuja plicata*
	Quercus alba	*Tilia cordata*
	Tilia americana	
	Tsuga heterophylla	
	Ulmus americana	

SOURCE: Reprinted, by permission, from Davies and Gerhold 1976, table 2.

Resistance to pollution may involve (1) avoidance of uptake of pollutants, (2) rapid incorporation of pollutants into less toxic products, together with dilution of the pollutant in a plant organ by extensive translocation, (3) biochemical resistance to the toxic substances, and (4) combinations of these.

Variations in stomatal characteristics sometimes are involved in pollution avoidance. Environmental factors influencing stomatal diffusion resistance often influence uptake of pollutants. For example, exposure of trees to SO_2 in the morning is more injurious than exposure in the afternoon, suggesting that leaf turgor and stomatal aperture (greater in the morning) are important factors controlling SO_2 uptake. Moisture stress before or during fumigation with pollutants decreases injury because of stomatal closure and reduced uptake of pollutants.

Stomatal size and frequency also affect pollution avoidance. *Fraxinus americana,* with large stomata (low stomatal diffusion resistance), absorbed more SO_2 than *Acer saccharum* with small stomata (high diffusion resistance) (Jensen and Kozlowski 1970). Experiments in our laboratory showed that some injury on exposure to SO_2 at low levels is associated with induction of stomatal opening or prevention of stomatal closure. Fumigation of *Ulmus americana* seedlings with 1 ppm of SO_2 for 8 hours induced stomatal opening presumably by effects on subsidiary cells. In contrast, fumigation with 2 ppm of SO_2 for 12 hours induced stomatal closure, probably reflecting injury to guard cells. Fumigation with 2 ppm SO_2 for 6.5 hours had a retarding effect on late-afternoon stomatal closure of *Fraxinus pennsylvanica* seedlings.

In some species pollution resistance appears to be complicated because it involves various degrees of pol-

TABLE 5
SUSCEPTIBILITY OF TREES TO OZONE

Sensitive	Intermediate	Resistant
Fraxinus americana	*Acer negundo*	*Abies balsamea*
Fraxinus pennsylvanica	*Cercis canadensis*	*Abies concolor*
Gleditsia triacanthos	*Larix leptolepis*	*Acer grandidentatum*
Juglans regia	*Libocedrus decurrens*	*Acer platanoides*
Larix decidua	*Liquidambar styraciflua*	*Acer rubrum*
Liriodendron tulipifera	*Pinus attenuata*	*Acer saccharum*
Pinus banksiana	*Pinus contorta*	*Betula pendula*
Pinus coulteri	*Pinus echinata*	*Cornus florida*
Pinus jeffreyi	*Pinus elliottii*	*Fagus sylvatica*
Pinus nigra	*Pinus lambertiana*	*Ilex opaca*
Pinus ponderosa	*Pinus rigida*	*Juglans nigra*
Pinus radiata	*Pinus strobus*	*Juniperus occidentalis*
Pinus taeda	*Pinus sylvestris*	*Nyssa sylvatica*
Pinus virginiana	*Quercus coccinea*	*Picea abies*
Platanus occidentalis	*Quercus palustris*	*Picea glauca*
Populus tremuloides	*Quercus velutina*	*Picea pungens*
Quercus alba	*Syringa vulgaris*	*Pinus resinosa*
Quercus gambelii	*Ulmus parvifolia*	*Pinus sabiniana*
		Pseudotsuga menziesii
		Quercus imbricaria
		Quercus macrocarpa
		Quercus robur
		Robinia pseudoacacia
		Sequoia sempervirens
		Sequoiadendron giganteum
		Thuja occidentalis
		Tilia americana
		Tilia cordata
		Tsuga canadensis

SOURCE: Reprinted, by permission, from Davies and Gerhold 1976, table 3.

lution avoidance and tolerance of pollutants. While resistant clones of *Picea abies* had more stomata per unit of leaf surface area than susceptible clones, gas exchange resistance was higher in the former (Braun 1977a). The resistant clones took up less SO_2 than the susceptible clones, and stomata of the former were more sensitive to environmental stresses (Braun 1977b). The data indicated that pollution resistance was associated largely with pollution avoidance rather than tolerance. Later studies (Braun 1977c, d), however, demonstrated that tolerance of pollutants also was significant: the resistant clones maintained a higher buffer capacity in unfumigated needles and were able to fix more sulfur in organic fractions, following SO_2 fumigation, than the susceptible clones.

One useful approach to studying the nature of resistance to pollution is by physiological and biochemical studies of clones of a given species known to differ in resistance to a given pollutant, or to a combination of pollutants. Because cell membranes are primary sites of action of oxidants, research is also needed on effects of individual and combined pollutants on cell ultrastructure and membrane permeability of species and genetic materials known to vary in resistance to pollution.

Water Supply and Tree Growth

BOTH THE DISTRIBUTION AND GROWTH OF TREES are controlled by water supply. Wherever trees grow, their development is limited by either too little or too much water, but mostly the former. Throughout much of the tree-supporting world many of the rains that fall do not recharge soils sufficiently with water to maintain maximum photosynthesis and growth for more than a short period after the rain. The water that evaporates from tree crowns, other plants, and the soil surface often exceeds the amount of water supplied during the growing season. This can only lead to physiological inefficiency and less than optimal tree growth. In many areas the extent of growth loss due to water stress is not realized because reliable data are not available to show how much more growth and wood production would occur if adequate moisture were available to trees during the entire growing season (Kozlowski 1968a).

SOURCES OF WATER

Trees obtain water primarily from the soil but they may also obtain some from the atmosphere and from adjacent trees if their roots are grafted together.

Atmospheric Moisture

A small amount of atmospheric water may be absorbed through leaves and stems. Under certain conditions atmospheric moisture appears to be ecologically significant. For example, in arid regions, where neither rain nor ground water is adequate to sustain development of naturally occurring vegetation, dew and fog appear to contribute appreciably to the water economy of plants. In hot deserts, dew is common during the night, and because stomata of most species are closed at night, leaves must absorb water through the cuticle. Evaporation of dew from leaves may decrease transpiration in the morning. Also, the absorption of dew during the night may reduce the high internal water deficits that develop in leaves during the day. Our experiments showed that presence of a heavy dew delayed the normal morning decline in water potential of *Populus* clones by two to three hours.

Absorption of atmospheric moisture is particularly well known in the highly hygroscopic shoots of salt-secreting shrubs. For example, when wilted *Tamarix* plants were placed in a saturated atmosphere they rehydrated and regained turgor (Waisel 1960). Uptake of appreciable atmospheric moisture in plants that do not secrete salts has also been documented. Waisel (1958) noted that during a prolonged drought in Jerusalem, dewed leaves of several species of shrubs and trees had a more favorable water balance early in the morning than undewed plants did. Duvdevani (1964) in Israel showed that plants deprived of dew grew appreciably less than those receiving dew. Gindel (1965) showed that several xerophytic woody plants could be successfully planted during the rainless season in Israel if water from dew and mist was collected and concentrated in the root regions. In the United States, young *Pinus ponderosa* plants in very dry soil survived for three weeks, but when their needles were sprayed at night with a fine mist they survived for seven weeks without additional watering of the soil (Stone 1957).

Some parts of California, Central America, the coast of Chile, the coast of eastern Mexico, and the Azores are places where conditions are suitable for collecting significant amounts of fog moisture. Went (1955) described vigorous growth of shrubs in "fog deserts" (areas with little precipitation—about one inch per year—but frequent fogs) in southwestern Africa and along the coast of Peru. In these regions, fogs form as the air rises against mountains lying near the coast. In the coastal lowlands, where there is no fog, there is practically no vegetation. By contrast, at an altitude of about 1,000 feet, where fog hangs most of the year, a very lush shrubby vegetation is present.

There is some very good evidence that atmospheric moisture is important in the water economy of trees in the Pacific Northwest. This is particularly true in the northern California fog belt where Azevedo and Morgan (1974) collected more than 40 cm of fog water under *Pseudotsuga* crowns. Dew also is ecologically important in areas of low rainfall with soils of low water-holding capacity. Fritschen and Doraiswamy

(1973) showed that accumulation of dew amounted to about 15 to 20 percent of the water that was evaporated from a 28-meter-high *Pseudotsuga menziesii* tree.

It has been suggested that in arid regions where humidity is high at night and the soil is very dry, water might move downward through the plant and possibly into the soil. However, Slatyer (1957) observed that when soil was very dry, water was absorbed from the air and accumulated in shoots. Although the diffusion gradient favored movement of water downward through the plant and into the soil, there was no evidence of such movement, presumably because of a discontinuity of contact between roots and soil. Vaadia and Waisel (1963) emphasized that xerophytes have thick cuticles and because of their leaf structure they are less well adapted than mesophytes for rapid foliar absorption of water at night. Experiments with tritiated water confirmed that water not only entered leaves of plants but increased leaf hydration and probably contributed to plant survival. However, because of high cuticular resistance, entry of water into leaves was slow and its further downward translocation would depend on reversal of the usual water potential gradients. This would require a much longer time than the short period during the night when dew is absorbed by leaves.

Water Transfer through Root Grafts

Root grafting among forest trees has very important physiological implications because transfer of water, carbohydrates, minerals, hormones, silvicides, microorganisms, and fungus spores often occurs from one tree to another through root grafts.

Natural root grafting among compatible forest trees is very common (Kozlowski and Cooley 1961). Root grafting leads to vascular connections brought about by union of cambium, xylem, and phloem of adjacent trees.

Eis (1972) presented evidence of extensive root grafting in *Pseudotsuga menziesii* stands, sometimes between trees 10 meters (m) apart. Suppressed trees apparently derived water and carbohydrates from the dominant trees to which they were grafted. This relationship may have postponed death of grafted suppressed trees in dense stands.

Bormann's (1966) data showed that relations between *Pinus strobus* trees connected by root grafts were influenced more by translocation of carbohydrates than water and minerals. Both water and minerals moved in the xylem and had a tendency to follow the grain of the wood to the crown. As xylem transport involves cross-grain movement, trees connected by root grafts did not appear to be able to divert very large amounts of water and minerals from trees to which they were grafted. However, quantitative data on amounts of water that can be obtained by trees through root grafts are lacking and much more research is needed to obtain such data for different species, sites, and crown classes.

Water Reservoirs within Trees

Transpirational demands can be fulfilled at least in part by water stored in various tissues. This is particularly evident in orchard trees having fruits with large amounts of water. Often high transpirational losses from leaves during the day lead to extraction of water from reproductive tissues while absorption from the soil is inadequate to supply the leaves.

In Calamondin orange trees movement of water from fruits to leaves on excised branches was shown by the higher percentage of moisture content of leaves on branches bearing fruits than of leaves on branches without fruits. Leaves on branches without fruits wilted faster than leaves on fruit-bearing branches (Chaney and Kozlowski, 1971). In lemon trees changes in fruit dimensions were sensitive indicators of variations in water content of leaves as it was influenced by the amount of moisture in the soil and climatic conditions (Bartholomew 1926). Furr and Taylor (1939) showed that diurnal water deficits occurred regularly in lemon trees, even when the soil was wet, and that transpiration induced translocation of water from fruits to leaves. Bartholomew (1926) noted that when leaves and lemon fruits were both attached to a branch, the leaves wilted more slowly than when fruits were removed from the branch.

In forest trees transpirational demands can be fulfilled, at least in part by water stored in foliage and in the stem near the cambium. The living cambial cells, undifferentiated xylem derivatives, phloem cells, and sapwood comprise an important water reservoir in the stem and woody roots. The availability of such reserves is emphasized by the large time lag involved in propagation of water potential (ψ) from transpiring leaves to the roots. Stems of severely water deficient *Acer saccharum* and *Betula papyrifera* trees shrank less per bar of daily ψ depression than those of trees with moderate water deficits. This difference was attributed in part to variations in amounts of sapwood water reserves (Pereira and Kozlowski 1978).

Water is withdrawn from tissues nearest to the site of evaporation. When such reserves are used those lower down on the stem are depleted. Recharge of stems occurs in a reverse sequence and, under favorable conditions, stems refill with water overnight.

In seedlings, water reserves in roots are important in elimination of daily water deficits. In large trees, however, stem reserves of water are more important than root reserves because too much time is involved in moving water from the roots to the leaves. Waring and Running (1976) estimated that in a mature *Pseudotsuga menziesii* forest there is enough water in the sapwood to supply a ten-day requirement for transpiration. They also estimated that the stemwood of one 80 m high *Pseudotsuga* tree contained more than 4,000 liters of water. The sapwood of the branches, because of its greater density, was estimated to contain only half the amount of water held by a comparable volume of stemwood.

Measurement of Water Deficits

Over the years water deficits in trees have been variously characterized and quantified in terms of moisture content (as percent of dry or fresh weight), relative water content (RWC) which has also been called relative turgidity, saturation deficit (SD), and water potential (ψ).

Moisture content of tissues, although easily determined and widely used, has limitations for comparing water status of different plants and tissues. Dry weight changes of tissues are not necessarily proportional to changes in the amount of water in tissues. Hence variations in moisture content (as percent of dry weight) over time do not always indicate changes in protoplasmic hydration or physiological activity. Dry weight changes of tissues often are the result of photosynthesis, respiration, and translocation. In rapidly growing leaves cell wall thickening accounts for rapid dry weight increases. Hence changes in leaf moisture content (as percent of dry weight) with time may reflect changes in hydration of physiologically active tissues, dry weight increment, or both. Seasonal decreases in moisture content of leaves of forest trees often are more the result of increases in leaf dry weight than dehydration of physiologically active cells (Kozlowski and Clausen 1965).

Expressing moisture content on a fresh weight basis also is unsatisfactory because the fresh weight of tissues such as leaves and young stems can vary widely, both between and within days. In addition, large changes in moisture content per unit of tissue result in very small changes in percentage of fresh weight (Kramer and Kozlowski 1960).

Saturation deficit (SD) and relative water content (RWC) compare moisture content of tissue at a particular time with moisture content of the same tissue when it is fully hydrated. They are calculated as follows:

$$SD(\%) = \frac{\text{saturated weight} - \text{original weight}}{\text{saturated weight} - \text{ovendry weight}} \times 100$$

$$RWC(\%) = \frac{\text{original fresh weight} - \text{dry weight}}{\text{fresh weight (fully turgid tissue)} - \text{dry weight}} \times 100$$

For comparative purposes both SD and RWC have serious limitations because leaves of one species at a certain moisture content may be fully turgid, whereas leaves of another species may be wilted at the same moisture content. This problem also applies to leaves of different ages on the same plant.

The most widely used measure of plant water deficit is water potential (ψ): the difference in chemical potential of water in a system and of pure free water at the same temperature. Water subjected to molecular restraints does not enter into physiological reactions within plants as readily as pure free water. Restraints may result from differences in pressure, salt concentration, adsorption at colloidal interfaces, confinement in capillaries, or inadequate water supply at a particular place.

The important components of ψ are:

1. Solute potential (ψ_s). The osmotic component reduces the chemical free energy of a solution as a function of the presence of dissolved salts, sugars, and other solutes.

2. Matric potential (ψ_m). The matric component reduces ψ as a function of capillary or colloidal forces by soil colloids, cell colloids, and cell walls. The force of adsorption between the matrix surface and the water molecules reduces ψ below that of pure free water.

3. Pressure potential (ψ_p). The pressure component may increase or decrease ψ depending on whether the molecules are subjected to pressures above or below atmospheric pressure. Under atmospheric pressure, the effect in an open system (e.g., soil) is zero. Turgor pressure in plants adds free energy to the system and ψ is increased. At wilting, the pressure component approaches or reaches zero and does not appreciably influence ψ.

In addition to ψ_s, ψ_m, and ψ_p, temperature and gravity affect total ψ.

In cells ψ, the sum of ψ_s, ψ_m, and ψ_p has a negative value, except in fully turgid cells when it is zero. As water deficits in plant tissues increase during droughts, their ψ values become increasingly more negative.

The many problems involved in characterizing and measuring water deficits in trees are discussed in the excellent chapter by Barrs (1968).

DEVELOPMENT OF WATER DEFICITS IN TREES

The path of water transport in trees includes movement in soil toward the roots, absorption by roots, movement across root tissues into xylem elements and upward to the leaves, evaporation into the intercellular spaces of the leaves, and diffusion through the stomata into the external atmosphere.

Water movement through the soil-plant-atmosphere system occurs along a path of decreasing potential energy. There is resistance to flow throughout the soil-plant-air continuum. The resistance is greater in the soil than in the tree, and is maximal in the transition from the leaves to the atmosphere where water changes from liquid to vapor. Total leaf resistance consists of stomatal, mesophyll, and cuticular components. Stomatal resistances are especially important because they are the only ones the tree can control. In stems of woody plants, water transport occurs in the sapwood, and resistance in the heartwood is very great. In gymnosperms, this resistance occurs through sealing of pit pairs through aspiration, occlusion with extractives, incrustation of pits with ligno-complex substances, and combinations of these. In angiosperms, resistance is provided by tyloses in vessels and various extractives.

In order for water to move upward through the soil-plant-air continuum the ψ_{plant} must be lower than ψ_{soil}, with the lowest ψ in the leaves. Transpirational water loss from leaves of a plant growing in wet soil progressively reduces the soil water content and ψ_{soil} (fig. 5). Associated with these changes is a reduction in ψ_{plant} (and water content of the plant), resulting in an increased internal water deficit. Hence, on a day-to-day basis, there is an overall decline in ψ of a plant growing in drying soil. Superimposed on this trend are diurnal variations in the internal water balance of plants. Basically these are controlled by relative rates of absorption of water and transpirational losses. During the day, transpiration exceeds absorption. The resulting internal water deficits in plants are reduced or eliminated during the night, when both absorption and transpiration are low, but absorption is greater.

Figure 5 shows changes in leaf water potential (ψ_{leaf}), root surface water potential (ψ_{root}), and soil mass water potential (ψ_{soil}) as transpiration occurs over a five-day period. During each day of the developing drought, transpirational water loss reduces ψ_{leaf}, and because absorption does not keep pace with transpiration, the internal water deficit increases until absorption equals transpiration. The internal water deficit is reduced only when absorption exceeds transpiration. When the soil is wet, water flow is sustained by small differences in ($\psi_{soil} - \psi_{root}$). When transpiration declines late in the day and during the night, plant water content increases so $\psi_{leaf} = \psi_{root}$, and by early morning $\psi_{leaf} = \psi_{root} = \psi_{soil}$. However, as ψ_{soil} decreases (as during days 3 and 4) higher values of ($\psi_{soil} - \psi_{root}$) are needed to maintain flow because the hydraulic conductivity of the soil decreases rapidly. By the fourth day, some stomatal closure decreases flow. Nevertheless, equilibration of ψ of the leaf, root, and soil occurs more slowly.

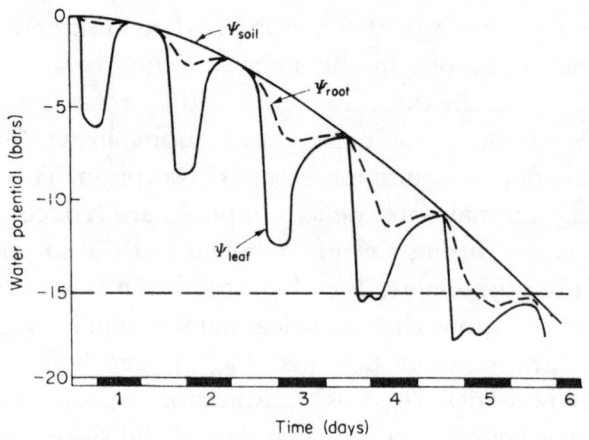

Fig. 5. Changes in leaf water potential (ψ_{leaf}), root surface water potential (ψ_{root}), and soil mass water potential (ψ_{soil}) as water is lost from a plant initially growing in wet soil. The horizontal dashed line represents the ψ_{leaf} at which wilting occurs. Reprinted, by permission, from Slatyer 1967, fig. 9.1. © 1967 by Academic Press, Inc. (London) Ltd.

RESPONSES OF TREES TO WATER DEFICITS

Water deficits in trees affect many physiological processes adversely and this results in loss of growth. In addition severe water stresses may injure trees and often kill them. Some of the responses of forest trees to water deficits will be discussed briefly.

Wilting

A direct effect of leaf desiccation is decrease in turgor which eventually causes wilting. However, the amount of water that must be lost to cause visible wilting of leaves varies for different species. Wilting is associated with a rather definite value of leaf water deficit in leaves of plants of a given species and age.

Wilting usually is classified as incipient, temporary, or permanent. Incipient wilting, which is characterized by slight decrease of turgor, usually does not cause drooping of leaves and occurs whenever conditions favor transpiration. Incipient wilting grades into temporary wilting, which is characterized by visible drooping of leaves during the day followed by rehydration and recovery from wilting during the night. During sustained periods of soil drying, temporary wilting grades into permanent wilting in which plants do not recover turgidity at night. Permanently wilted plants can recover turgidity only when water is added to the soil. Prolonged permanent wilting usually kills most species of plants.

The relation between water loss from leaves and visible wilting is complicated by large differences among species in the amount of supporting tissues their leaves contain. Leaves of *Prunus serotina*, *Cornus* spp., and delicate-leaved shrubs wilt readily. Rhododendrons are very sensitive to drought, and their leaves curl prior to yellowing and turning brown during a drought. The leaves of *Syringa vulgaris* warp and wrinkle, and then they yellow, turn brown, and often are shed early as a result of water shortage. By comparison, the leaves of *Ilex* and *Pinus* are permeated with abundant sclerenchyma tissue and do not droop readily even after they lose considerable water.

PHYSIOLOGICAL PROCESSES

Internal water deficits affect a number of physiological processes sequentially and often concurrently. An early response to plant water deficit is a decrease in cell enlargement caused by reduction in ψ. This is followed by reduced cell wall and protein synthesis in tissues with high growth potential. As tissue water deficits increase further, cell division may be inhibited and amounts of some enzymes decline. Stomatal closure begins and, consequently, rates of photosynthesis and transpiration decrease. Abscisic acid (ABA) begins to accumulate. By this time numerous secondary and tertiary changes occur. Further desiccation is accompanied by substantial decreases in respiration and translocation of carbohydrates and growth regulating hormones. Amounts of some hydrolytic enzymes may increase and ion transport decreases. Finally, proline accumulates and the rate of photosynthesis becomes low or negligible. Leaf senescence becomes apparent and old leaves are shed. These changes inevitably become associated with further reduction in plant growth (Hsiao 1973). For more details on effects of water deficits on various plant processes of healthy and diseased herbaceous and woody plants the reader is referred to *Water Deficits and Plant Growth,* vols. 1–5 (Kozlowski 1968a, 1968b, 1972c, 1976b, 1978).

Stomatal Aperture. One of the earliest responses in leaves to a mild water stress is stomatal closure. Stomata often close during early stages of a drought, often long before leaves wilt. In a given plant the time of stomatal closure may vary with leaf size and stomatal size. Stomata of shade leaves are more sensitive than those of sun leaves to water stress. The stomata of young leaves tend to close more rapidly than those of old leaves in response to drought conditions.

Many forest trees close stomata temporarily in the middle of the day in response to rapid water loss. Midday stomatal closure generally is followed by reopening and increased transpiration in the late afternoon before the final daily closure occurs as light intensity decreases. The extent of midday stomatal closure depends on air humidity and soil moisture availability. One study showed that when soil moisture was readily available, about 70 to 100 percent of the stomata of apple trees opened in the morning. They usually remained open throughout the day when air humidity was high, even though the soil had dried considerably. However, when the soil was charged with water the stomata closed before noon if temperature was high and humidity low. As soil dried, the daily duration of stomatal opening was reduced. When the soil was very dry (e.g., at or near wilting percentage) the stomata tended not to open at all. If the air was nearly saturated, the stomata opened for a short time but usually closed within an hour (Magness, Degman, and Furr 1955).

Water loss from plants is prevented by successively

earlier stomatal closing during each day of a developing drought as well as by temporary stomatal closure during the middle of the day. However, stomatal closure during a drought may not prevent killing of plants that lose appreciable amounts of water directly through the leaf epidermis after the stomata close.

Some plants close stomata early during developing droughts and others do not. Gymnosperms usually undergo more leaf dehydration than angiosperms before they close their stomata. One of the unfortunate effects of severe drought is that it sometimes causes permanent damage, and stomata of some species may open slowly or not at all when the plants are rewatered. Under these conditions the leaves may recover from wilting, but the stomata may not reopen after irrigation. Both the severity of a drought and species influence subsequent photosynthetic recovery. Zavitkovski and Ferrell (1970) found that the rate of photosynthesis of *Pseudotsuga menziesii* trees that had been subjected to drought remained low even after the stressed trees were rewatered.

Diurnal Variations in Stomatal Aperture. Two generalized patterns have been reported. Usually stomata open in the morning and gradually close during the day or they may close rapidly rather early in the day and remain closed until the following day.

Stomatal aperture of well-illuminated leaves is influenced simultaneously by several external and internal variables including air temperature, vapor pressure deficit, ambient CO_2 concentration, and leaf water potential. Hence, when stomatal aperture is plotted against one of these variables a scatter diagram results

(Jarvis 1976). We have found with both gymnosperms and angiosperms that, when internal water stresses are low, stomatal aperture is closely related to photosynthetically active radiation. During droughts, however, stomata close readily during the day when light intensity is high. Examples of diurnal changes in stomatal aperture and environmental factors at various times during the growing season are shown in Figures 6 and 7.

Stomatal closure is induced by dehydration of leaves or dehydration of guard cells and adjoining epidermal cells. The sensitivity of stomata to leaf water stress apparently varies considerably among species. Stomata of pine appear to be more sensitive than those of other gymnosperms. Lopushinsky (1969) reported that stomatal closure occurred at the following values of leaf ψ: *Pinus contorta*, -14.6 bars; *Pinus ponderosa*, -16.5 bars; *Picea engelmannii*, -16.0 bars; *Pseudotsuga menziesii*, -19.0 bars; and *Abies grandis*, -25.1 bars. In *Abies balsamea* and *Pinus resinosa* leaf ψ near -18 bars induced stomatal closure (Pereira and Kozlowski 1976). Our experiments showed that young *Pinus banksiana* trees avoided drought better than *P. resinosa* trees because the former maintained a higher needle ψ during most of the growing season. This reflected lower transpiration and slower water depletion associated with lower leaf area and stomatal closure at a higher needle ψ in *Pinus banksiana* (Pereira and Kozlowski 1977a). Critical values of ψ for stomatal closure also vary considerably among angiosperms. Threshold values of -15 bars for *Acer circinatum* and -25 bars for *Quercus alba* have been reported (Lassoie and Scott

1977; Hinckley and Bruckerhoff 1975). Very much lower values have been reported for desert shrubs.

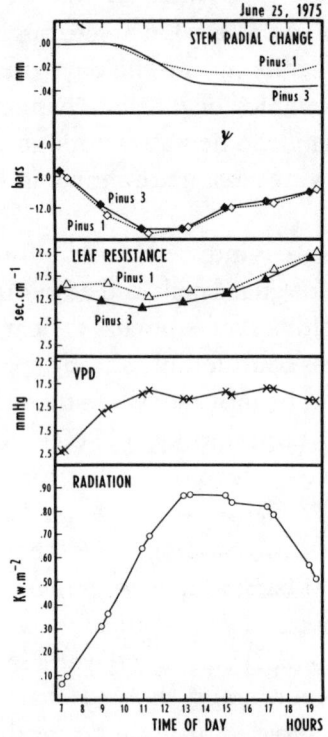

Fig. 6. Typical diurnal changes in *Pinus resinosa* of stem shrinkage, leaf water potential (ψ), stomatal aperture as shown by leaf resistance (increasing leaf resistance indicates stomatal closure), vapor pressure deficit (VPD), and solar radiation. Reprinted, by permission, from Pereira and Kozlowski 1976, fig. 8.

Stomata of adaxial and abaxial leaf surfaces may also respond differently to critical values of leaf ψ. In *Eucalyptus camaldulensis* seedlings stomata of the abaxial epidermis closed gradually at ψ values between -8 and -12 bars. Those of the adaxial epidermis closed rapidly at ψ values below -9 bars (Pereira and Kozlowski 1976). As leaves age, stomata become less responsive and stomatal resistance in the middle of the day may increase appreciably because stomata do not open completely (Slatyer and Bierhuizen 1964).

Shrinking and Swelling of Plant Tissues and Organs. Variations in expansion and contraction of vegetative and reproductive tissues as a result of changes in hydration are well known. Various tissues and organs of trees shrink during the day, when transpiration exceeds absorption, and they expand at night as trees tend to rehydrate. A few examples of seasonal and diurnal shrinkage and expansion of leaves, stems, roots, and reproductive structures will be given.

Daily shrinkage and expansion of leaves of English Morello cherry (*Prunus cerasus* grafted on *Prunus mahaleb* rootstock) were related to environmental conditions affecting stomatal aperture and transpiration. As early morning increases in vapor pressure deficit (VPD) occurred, the leaves shrank rapidly and, when VPD decreased, the leaves expanded. The negative correlation between VPD and leaf shrinkage was much better when soil moisture content was high than when it was low (Chaney and Kozlowski 1969a). Negative correlations between VPD and leaf thickness also were shown for Calamondin orange (Chaney and Kozlowski 1971). Leaf thickness began to decrease near sunrise, at about the time VPD began to increase or slightly later, and the decrease continued to mid- or

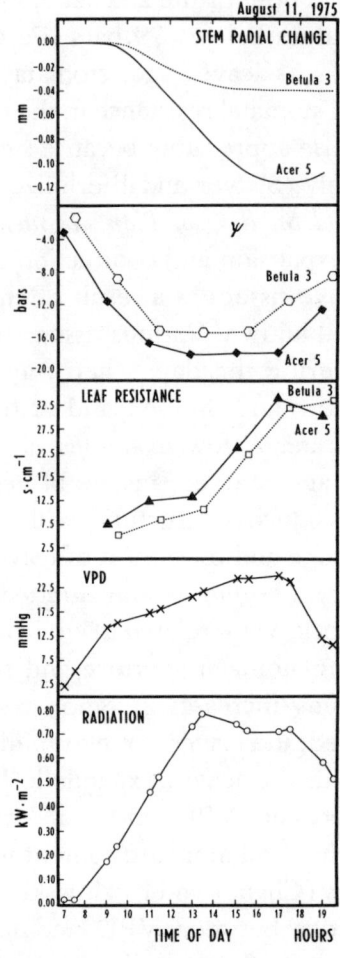

Fig. 7. Typical diurnal changes in *Betula papyrifera* and *Acer saccharum* of stem shrinkage, leaf water potential (ψ), stomatal aperture as shown by leaf resistance (increasing leaf resistance indicates stomatal closure), vapor pressure deficit (VPD), and solar radiation. Reprinted, by permission, from Pereira and Kozlowski 1978, fig.5.

late afternoon, by which time VPD was decreasing. Parker (1952) showed considerable shrinkage of pine needles during developing droughts.

Because of recurrent dehydration and rehydration tree stems usually shrink slightly during the day and they expand at night (Kozlowski 1972b). Kozlowski and Winget (1964) found that amounts of daily shrinking of tree stems in Wisconsin varied greatly during the summer, with small amounts occurring early in the growing season, followed by increased shrinkage in midseason, and greatly decreased shrinkage in late summer after transpiration had depleted the soil and tree reserves of water.

In addition to reversible diurnal stem shrinkage, seasonal shrinkage of stems occurs commonly during droughts. The radial decrease of stems during droughts often exceeds the amount of radial increase as a result of cambial growth during the same period. For example, many stems of *Pinus strobus* trees in New England showed net weekly radial decreases during the summer (Bormann and Kozlowski 1962). Dimock (1964) reported that *Pseudotsuga menziesii* stems shrank consistently during a six-week period in the summer. Buell, Small, and Monk (1961) demonstrated that during a severe drought in New Jersey in August 1957 tree stems shrank so much that their diameters were smaller than they were before the growing season started. When internal water balance was finally restored by rains in December the tree stems expanded rapidly. Kozlowski, Winget, and Torrie (1962) recorded marked swelling of tree stems on each day following rain during the summer.

Braekke and Kozlowski (1975) showed that both diurnal and seasonal shrinkage and swelling of stems were greater in *Pinus resinosa* than in *Betula papyrifera*. Stems of some *Pinus* trees shrank progressively after cessation of seasonal cambial growth. *Betula* trees did not show late-season stem shrinkage or appreciable diurnal shrinkage and swelling after cambial growth stopped.

Generally more diurnal shrinkage occurs in upper parts of tree stems than near the base. In *Pseudotsuga menziesii* the amount of diurnal stem shrinkage increased from the stem apex to a point near mid-crown and then decreased toward the base of the stem (Dobbs and Scott 1971).

Since water stress in root tissues follows its development in stems, shrinking and swelling of roots may also be expected. In *Pinus radiata* woody roots shrank during the day and swelled during the night (MacDougal 1936).

As mentioned, fruits often act as water reservoirs. Diurnal expansion and contraction have been reported for a variety of angiosperm fruits including acorns, apples, cherries, oranges, lemons, peaches, plums, walnuts, pears, and avocados (Kozlowski 1972b). Young fruits often show less diurnal shrinkage than old fruits. This sometimes is correlated with a low moisture content in the young fruits or a low transpiring leaf area on the plant bearing them.

Both diurnal and seasonal shrinkage of gymnosperm cones have been reported. Early in their development cones progressively increase in diameter and show little, if any, superimposed midday shrinkage. In a mid-stage of development, cones usually show recurrent shrinkage during the day and swelling during the night. During late stages of their development gymnosperm cones usually show predominant continuous shrinkage (Dickmann and Kozlowski 1969; Chaney and Kozlowski 1969b).

Internal water deficits inhibit photosynthesis by inducing closure of stomata as well as changes in chloroplasts. The importance of stomatal control is shown by correlations between stomatal closing and decrease in photosynthesis during droughts. Photosynthesis also appears to be inhibited by changes in cell metabolism during drought. Boyer (1973) showed that there was a general inhibition of the light reactions of photosynthesis when plants were under water stress. The combined loss in leaf area and photosynthetic activity represents a potentially large loss of photosynthate for forest trees.

The rate of photosynthesis is correlated with stomatal aperture. When stomata begin to close, photosynthesis decreases, often early in a soil drying cycle. Seasonal cycles of photosynthesis in trees are closely related to availability of soil moisture as it affects stomatal aperture. The rate of photosynthesis often declines in the middle of the day because of stomatal closure at that time (Kozlowski and Keller 1966).

There has been some disagreement about whether or not photosynthesis is reduced significantly in drying soil before the wilting percentage is reached. Some of the disagreement stems from the fact that investigators who measured photosynthesis and soil moisture content at the same time often assumed that soil water deficits and internal water deficits in leaves usually were proportional. However, trees growing in dry soil may not develop high internal water deficits if atmospheric conditions are conducive to low transpiration (e.g., when relative humidity of the air is high). Conversely, during periods of high temperature and low relative humidity, even trees in nearly saturated soil may undergo severe water stress because of excess transpiration over absorption of water. This reemphasizes that internal water deficits depend on relative rates of absorption and transpiration and not on absorption of water alone.

Leaf Shedding

Water deficits often induce leaf shedding by trees. In some species this involves true abscission; in others the leaves simply wither and decay.

Shedding of leaves during dry summers occurs commonly, with the reduction of leaf surface varying with species and with moisture conditions from year to year. Many forest trees drop all or almost all of their leaves during a severe drought, whereas *Eucalyptus* and *Citrus* tend to shed only some of their leaves.

In broad-leaved trees, early leaf fall and injury in response to drought occur commonly. For example, leaf fall of deciduous street trees occurred in the United States during the severe 1913 drought. In some cases, almost all leaves were shed by the end of

July. Near Lincoln, Nebraska, *Celtis occidentalis, Ulmus americana,* and *Populus deltoides* were conspicuously affected. Toward the end of the summer a number of trees, that had been defoliated by drought, put out a second crop of small leaves from previously dormant buds. Most conspicuous examples were *Celtis* and *Gymnocladus dioicus* (Pool 1913).

The twelve-month period following June 1933 was the driest recorded for the Dakotas, Minnesota, Nebraska, Iowa, Illinois, and Missouri. Kansas, Oklahoma, and Colorado also were undergoing extreme drought conditions (Kincer 1934). In Nebraska, many trees were partially defoliated by drought early in the summer of 1934. Among the species that showed extensive leaf scorching and defoliation were *Prunus virginiana, Salix interior, Ulmus americana, Quercus borealis maxima,* and *Tilia americana.* Injury to foliage and defoliation were most apparent in portions of the crown that were in full sun. By the end of August, leaf rolling, folding, curling, and shedding were intensified (Albertson and Weaver 1945).

Often the shedding of leaves is associated with hot, dry winds. In mountainous regions, such as the east slope of the northern and central Rocky Mountains, warm, dry chinook winds appear suddenly. In southern California, hot, dry "northers" from the interior deserts cause extensive desiccation injury to citrus groves.

In California, leaf desiccation of *Citrus* by dry winds is followed by two types of injury described as "windburn" and "scorch" (Reed and Bartholomew 1930).

Leaves killed by windburn first wilt, then dry out rapidly, and become brittle within a day. If the wind stops after a few hours, the wilted leaves sometimes recover. However, if the leaves are desiccated beyond a critical threshold, they do not recover. Old leaves are killed faster than young ones. In a few days, the windburn-killed leaves are shed. Scorch, the less common type of injury, results from hot winds. Blades of scorched leaves generally turn brown and become brittle within a few hours, without an intermediate wilting stage. Scorched leaves often remained attached to twigs for several weeks, even though exposed to strong winds.

Many trees of tropical rain forests lose all their leaves in response to even mild droughts, with their pattern of leaf shedding not necessarily tied to an annual cycle. Richards (1952) considered variations among rain forest trees to be so great that he could not distinguish clearly between evergreen and deciduous trees. For convenience, he regarded a species as evergreen if its members carried a substantial number of leaves throughout the year. Deciduous species were considered to be those that lost all or almost all of their leaves, even if only for a few days. Deciduous trees, as defined above, are numerous in rain forests in all tropical areas, even in climates with very evenly distributed rainfall.

In tropical areas with a distinct dry season, as in the "caatinga" and "cerrado" of Brazil, much leaf shedding occurs at the height of the dry season. Individuals of some species lose all their leaves and those of other species shed only some of their leaves before new ones grow. Hence, the mass of woody plants has only a fraction of the living leaves that were present in the wet season. In many species, new leaves emerge even before the rains start again. The leaves that are retained during the dry season lose considerable water by stomatal transpiration. Such transpiration is possible because the plants have deep roots that obtain water from moist soil above a deep water table (Eiten 1972).

Control of Leaf Shedding. When true abscission occurs under conditions of water stress, important internal changes occur in hormone balance and in synthesis and activity of enzymes which eventually hydrolyze pectins of the middle lamella between cells of the abscission layer.

Several hormones are involved in inducing changes leading to abscission, but ethylene appears to be most important. The auxin gradient across the abscission zone exerts a strong regulatory effect on leaf separation. Auxin moves from the leaf to the abscission zone and there maintains physiological processes so as to delay abscission. Auxin also has a role in mobilizing nutrients from weak to vigorous organs and thereby promotes abscission of the weaker organs. Abscisic acid (ABA) appears to accelerate preabscission senescence changes in tissue distal to the abscission zone. Gibberellin tends to inhibit abscission by promoting growth of the subtending organ. However, when it is applied to abscission zones proximal to them it induces abscission. In part, gibberellin may also act by promoting synthesis of auxins or hydrolytic enzymes. The influences of cytokinins are similar to those of auxin in maintaining biochemical processes and directing flow of nutrients. These effects may either retard or promote abscission (Addicott 1970).

Ethylene is a powerful inducer of abscission, with concentrations as low as 0.1 μl/liter capable of initiating the process. In most species where ethylene production was monitored, tissues distal to abscission zones produced large quantities of ethylene during senescence of the tissue and prior to abscission. Applied ethylene or large amounts of ethylene produced by leaves of evergreens that had been treated with auxin induced senescence and abscission of leaf blades and petioles (Hallaway and Osborne 1969). The capacity of ethylene to induce and accelerate abscission depends on the level of auxin at or near the abscission layer.

In normal abscission, blade senescence involving loss of chloroplyll precedes the actual shedding of leaves. Following a severe drought, however, leaves often are shed while still somewhat green. Sometimes leaf shedding of water-deficient plants may not occur until after rehydration following drought damage that is not immediately apparent. Then leaves fall rapidly, suggesting that abscission is initiated by a response to water stress injury that cannot be completed without adequate water. This is in accord with the observation that water is needed for normal functioning of cells of the abscission zone. Hence it appears that, in some species undergoing water deficits, the leaf abscission zone cannot compete successfully with the rest of the plant for water or obtain enough water for hydrolysis of the middle lamella and cell walls. Also, rehydration allows stems to regain turgor and, thereby, to facilitate shedding of partly abscised leaves (Addicott and Lynch 1955).

Growth Inhibition

Growth of vegetative and reproductive tissues is inhibited by reduction in cell enlargement as well as cell division, with the former usually more sensitive to water stress.

Shoot Growth. Internal water deficits in trees impede growth of shoots by influencing development of shoot primordia and by subsequently suppressing expansion of these primordia. Thus a period of drought has a carry-over effect in many species from the year of bud formation to the year of expansion of the bud into a shoot. Drought also has a short-term effect in inhibiting expansion of shoots within any one year (Kozlowski 1971a). These effects of drought will be considered separately.

Water Deficits and Shoot Growth in the Same Year. The shoot growth response to water deficit varies with species, severity, and time of occurrence of the drought, and even with location of shoots on a tree. Internode expansion and leaf expansion also may be affected to different degrees by water deficits.

To illustrate the variable response of different species to a late-season drought in the United States (e.g., in late July or early August), it is important to recognize that species vary greatly in seasonal duration of shoot growth. Shoots of some species elongate in only two to six weeks, whereas those of other species expand over a period of several months.

Whereas a late July or early August drought may not affect current-year shoot elongation of species exhibiting fixed growth, whose shoots expand only during the very early part of the frost-free season, it will, in contrast, inhibit expansion of shoots such as heterophyllous and recurrently flushing species which elongate their shoots during much of the summer. But the situation is not that simple because predetermined species which *normally* expand their shoots during the early part of the frost-free season only, sometimes produce some abnormal late-season lammas or proleptic shoots from opening and expansion of buds which normally would not open until the following year. Clearly the expansion of such abnormal late-season shoots can be affected by severe late-season water deficits. Another complication is that some shoots on a tree are affected differently by droughts than are other shoots. For example, in *Larix* both short and long shoots are produced. The short shoots consist mostly of needles that expand early, whereas the long shoots continue to expand both needles and internodes late into the summer. Therefore, a late summer drought may be expected to inhibit growth of long-shoot internodes and to have little effect on the short shoots which do not undergo internodal expansion and whose leaf expansion precedes the drought (Clausen and Kozlowski 1967). In the southern pines, late summer droughts may be expected to influence expansion of shoots of the upper crown to a greater extent than those of the lower crown. This is because the number of seasonal growth flushes varies with shoot location in the crown. Shoots in the upper crown exhibit more seasonal growth flushes than those in the lower crown. In fact, buds of some lower branches may not open at all (Kozlowski 1971a).

In some species the seasonal duration of leaf expansion is very different from that of internode expansion. In *Pinus strobus,* for example, shoot internodes expanded in only 35 to 40 days but needles required 85 to 95 days (Lister et al., 1967). Therefore, when internode expansion ceases early, a drought thereafter may affect leaf expansion but not internode expansion. Lister et al. (1967) compared shoot growth of *Pinus strobus* seedlings, maintained under water stress for much of the growing season, with shoot growth of well-watered seedlings. Water stress caused only a slight reduction of internode elongation and a proportionally greater decrease in needle elongation. Both the rate and duration of needle growth were decreased by drought.

Cambial Growth. Water deficits in trees play a major role in controlling cambial growth. The width of the annual ring and its distribution along the bole, seasonal duration of cambial growth, proportion of xylem to phloem increment, time of latewood initiation, and duration of latewood production are influenced by the amount of water and its availability at different times of the season (Kozlowski 1971b). Evidence of marked sensitivity of cambial growth to water supply comes from (1) correlations of xylem increment with rainfall or available soil water, (2) irrigation studies, and (3) thinning studies. The fact that the width of annual xylem rings varies from year to year with water supply has provided the basis for the science of dendrochronology.

As internal water deficits develop in trees during midsummer droughts, cambial growth slows or ceases and accelerates or resumes with the next rain. During a dry summer, diameter growth of *Pinus taeda* in Arkansas stopped by August. However, it resumed during September when considerable rain fell. About a third of the total season's xylem increment was produced during September and October (Zahner 1958).

Thinning often increases the rate of cambial growth of the residual trees in a stand. Such accelerated growth represents an integrated physiological response to more water, light, and minerals than were available prior to the thinning. The amount of increase in cambial growth often is proportional to the degree of thinning. Thinning also extends the seasonal duration of cambial growth. In Arkansas, for example, widely-spaced trees grew into late autumn, whereas closely spaced trees stopped growing in midsummer, by which time they had depleted the available soil moisture (Zahner and Whitmore 1960).

Sometimes oversevere thinning may set in motion conditions that lead to deterioration of growth of the residual trees. In Wisconsin, for example, Haberland and Wilde (1961) severely thinned dense *Pinus resinosa* stands. This was followed by invasion of weeds and eventually by reduction of available water to the trees. As a result, diameter growth of the residual trees was reduced. In such situations, frequent light thinnings may improve soil moisture conditions, whereas heavy thinnings may not.

Severe internal water deficits in trees often cause formation of discontinuous rings or sometimes prevent any xylem from forming in the lower stem. For example, Fritts et al. (1965) studied tree ring characteristics from the forest interior to a semi-desert forest border. As moisture became limited toward the dry forest border, the percentage of absent xylem rings increased sharply. These observations confirmed those of Glock, Studhalter, and Agerter (1960), who noted that annual rings of trees in the forest interior were of rather uniform thickness, whereas at the forest border they were made up of variable thick, thin, and partial xylem increments. In a forest border in the arid Southwest, as fluctuations in soil moisture became more intense and rapid, increasing numbers of partial layers of wood resulted. Cambial activity following a rain varied from production of a few isolated large xylem cells to incomplete growth layers and even complete, entire growth layers. The production of late-season lammas shoots in response to abundant rainfall often results in cambial reactivation and formation of false (multiple) rings of wood.

Lag Effects of Water Supply on Cambial Growth. In addition to controlling cambial growth while a drought is in progress, water stress may also be expected to affect cambial growth in the subsequent year, or even years. Such a lag effect is the result of the influence of water supply on crown development and its physiological activity.

Although total shoot elongation of species such as the southern pines (e.g., *Pinus taeda, P. echinata*) often involves formation and opening of several buds on a shoot within one season, the majority of temperate-zone species produce an annual shoot from expansion of a single bud only. Hence, shoot formation of the latter group is a two-year process involving bud differentiation the first year and extension of parts within the bud into a shoot during the second year (see previous section). As water deficits during the year of bud formation often regulate the number of leaf primordia laid down, they influence the number of leaves, and hence leaf surface, in the following year when the predetermined shoots expand.

Cambial growth, which depends on a downward flow of carbohydrates and hormonal growth regulators from the leaves, varies greatly with leaf development. And, as mentioned, leaf development depends to a considerable extent on prior-year weather. It is not surprising, therefore, that cambial growth is controlled by water supply of both the current and previous year. Zahner and Stage (1966) used regression analysis to study the influence of water deficits on diameter increase of *Pinus monticola* in northern Idaho. Droughts during both the previous and current growing season inhibited basal area increase, further emphasizing both rapid and greatly delayed responses of cambial growth to water deficits.

Although thinning of stands improves light and soil moisture conditions for residual trees, the beneficial effects on cambial growth sometimes are not transmitted to the lower stem for a long time. This is especially true for trees with sparse crowns which tend to concentrate xylem increment in the upper stem. When such trees are released by thinning, their increased production of carbohydrates and hormones often is directed first toward crown development. Only after a rather long lag period does cambial growth accelerate in the lower stem. For example, cambial growth of stagnated *Pinus ponderosa* trees was greatest at 80 percent of stem height and least at the base. After thinning, diameter growth at the base of stems of remaining trees was unaffected for two years during which time the tree crowns increased in size. In the third year after thinning, however, the widest growth rings were produced at the base of the stem (Myers 1963).

In addition to controlling the width of the annual ring, water supply influences the relationship between earlywood and latewood which, in turn, affects specific gravity and quality of wood. The availability of water affects the time of latewood initiation, the length of time during which latewood is produced, and the abruptness of the transition between earlywood and latewood. Drought triggers early formation of latewood and sustained subsequent drought shortens the period of latewood formation (Kozlowski 1971b).

Several studies showed that low water supply causes

latewood formation to start early in the growing season. For example, in 1954 soil moisture decline in Michigan began on June 21, and latewood formation in *Pinus resinosa* began at the end of June. During the next year, however, soil moisture declined early, between May 23 and June 29, and the transition to latewood also occurred early, between May 27 and June 9 (Kraus and Spurr 1961). Zahner and Oliver (1962) showed that *Pinus resinosa* trees released by thinning began forming latewood about two weeks later than trees in unthinned stands. The longer duration of earlywood formation in the released trees was ascribed to delay in depletion of soil moisture.

High water availability late in the season increases the total width of the latewood band. For example, irrigating *Pinus ponderosa* trees in eastern Washington resulted in as much as a 40 percent increase in latewood width (Howe 1968).

Reproductive Growth

Flowering and fruiting of trees are influenced by water supply at any stage during flower-bud formation, flowering, pollination, fertilization, embryo growth, or fruit and seed development. However, it is difficult to make broad generalizations about effects of drought on reproductive growth, partly because timing of the cycle of reproductive growth varies widely.

Formation of flower buds often is suppressed by drought, but the effect may not be obvious until the following year at flowering time. Sometimes, however, a mild water deficit at a critical time will increase initiation of flower buds by decreasing growth of vegetative tissues. Both the size and quality of fruits are improved by an adequate water supply before and during the period of fruit enlargement.

Drought Injury

Trees may be variously injured by drought depending on the severity and duration of internal water deficits. In deciduous trees curling of leaves, bending, mottling, marginal browning, "scorching," chlorosis, and early autumn coloration are well-known responses to drought. In conifers droughts may cause yellowing of needle tips, which subsequently turn brown. Discoloration of leaves as a result of excessive transpiration by evergreens occurs commonly. During some winter and spring days, when the air warms during the afternoon, needles of conifers lose water appreciably but water cannot be taken up by the roots from the cold or frozen soil and leaves consequently dry to a dangerous level. In the Adirondack Mountains of New York State entire mountainsides were covered with discolored conifers as a result of excessive evaporation of water from leaves during the winter. Such injury has sometimes been referred to as "red belt" or "parch blight."

As drought intensifies, its harmful effects may be expressed in dieback of twigs and branches. There are a number of complex "dieback-decline" diseases which are not fully understood but which often involve response to water deficit, at least in part. These diebacks have been reported in a number of species including *Fraxinus americana, Acer saccharum, Betula papyrifera, B. alleghaniensis, Liquidambar styraciflua, Quercus rubra, Q. velutina,* and *Fagus sylvatica.* Symptoms of such decline diseases include gradual or sometimes sudden dying of individual branches. These symptoms may continue for as long as several years, after which the tree may die.

If droughts are very severe and prolonged, the stems may become so dried out that they crack. Drought cracks are more common in gymnosperms than in angiosperms and *Picea* appears to be injured more than *Abies balsamea, Tsuga candensis, Larix* spp., or *Pseudotsuga menziesii.* In some species longitudinal flutes or hollows in tree stems, dying of bark in vertical strips, and longitudinal cracking of bark result from water deficits. In other species drought cracks go all the way to the pith. Depending on the extent of injury, drought cracks may or may not heal. If the wounds do not heal, decay fungi and insects often invade trees through the openings. Drought cracks sometimes are mistaken for frost cracks. However, drought cracks tend to be wider in the middle than at the ends and so do not heal as readily as frost cracks. Drought cracks may also be mistaken for lightning cracks, but the former usually are shorter. Lightning cracks sometimes run the full length of the stem.

DROUGHT RESISTANCE

Although both drought avoidance and desiccation tolerance (capacity of protoplasm to withstand severe desiccation) are involved in overall drought resistance, drought avoidance is much more important in survival of trees under extremely arid conditions (Kozlowski and Davies 1975).

Drought Avoidance

Drought-avoiding adaptations may be found in leaves, stems, or roots. Drought avoidance often involves more than one adaptation. Furthermore, adaptations for drought avoidance often vary for different species of trees growing side-by-side. Among the important adaptations for drought avoidance are shedding of leaves; production of small or only few leaves; small, few, and sunken stomata; rapid closure of stomata during drought; thick leaf waxes; and strong development of palisade mesophyll. Although the volume of intercellular spaces often is less in xeromorphic than in mesomorphic leaves, the ratio between the internal exposed surface of the leaf and its external surface area usually is higher in xeromorphic leaves. Stem adaptations for drought avoidance include capacity for twig and stem photosynthesis, and development of a wide cortex that protects vascular tissues from desiccation. The most important drought-avoiding adaptations of roots are capacity for extensive root growth (high root-shoot ratios), high root regenerating potential immediately after transplanting, and production of adventitious roots near the soil surface.

Stomatal Characteristics. The amount of resistance to water loss from leaves varies among species and genetic materials because of variations in stomatal size, stomatal frequency, and capacity for rapid stomatal closure during developing droughts. In many species stomatal size and frequency are negatively correlated. For example, whereas *Acer saccharum* and *A. saccharinum* had many small stomata, *Ginkgo biloba, Betula papyrifera,* and *Gleditsia triacanthos* had relatively few large stomata. Stomatal size and frequency among different species of the same genus may also vary greatly, as in species of *Quercus, Fraxinus,* and *Crataegus* (Davies et al. 1973). Variations in stomatal size and frequency in two *Populus* clones are shown in Figure 8.

We found that transpiration rates differed greatly among *Populus* clones and were better related to variations in stomatal aperture than to stomatal size and frequency or to such aspects of leaf anatomy as leaf thickness, epidermal thickness, amount of palisade parenchyma, amount of spongy parenchyma, percentage of combined thickness of spongy mesophyll plus epidermal layers, or percentage of palisade parenchyma (Siwecki and Kozlowski 1973).

Wide variations occur among species in stomatal responses to atmospheric factors, such as high temperature, high light intensity, low humidity, and wind, which induce water deficits in trees (Davies and Kozlowski 1974; Davies, Kozlowski, and Pereira 1974). For example, experiments were conducted on species that occur along an ecological gradient from xeric to

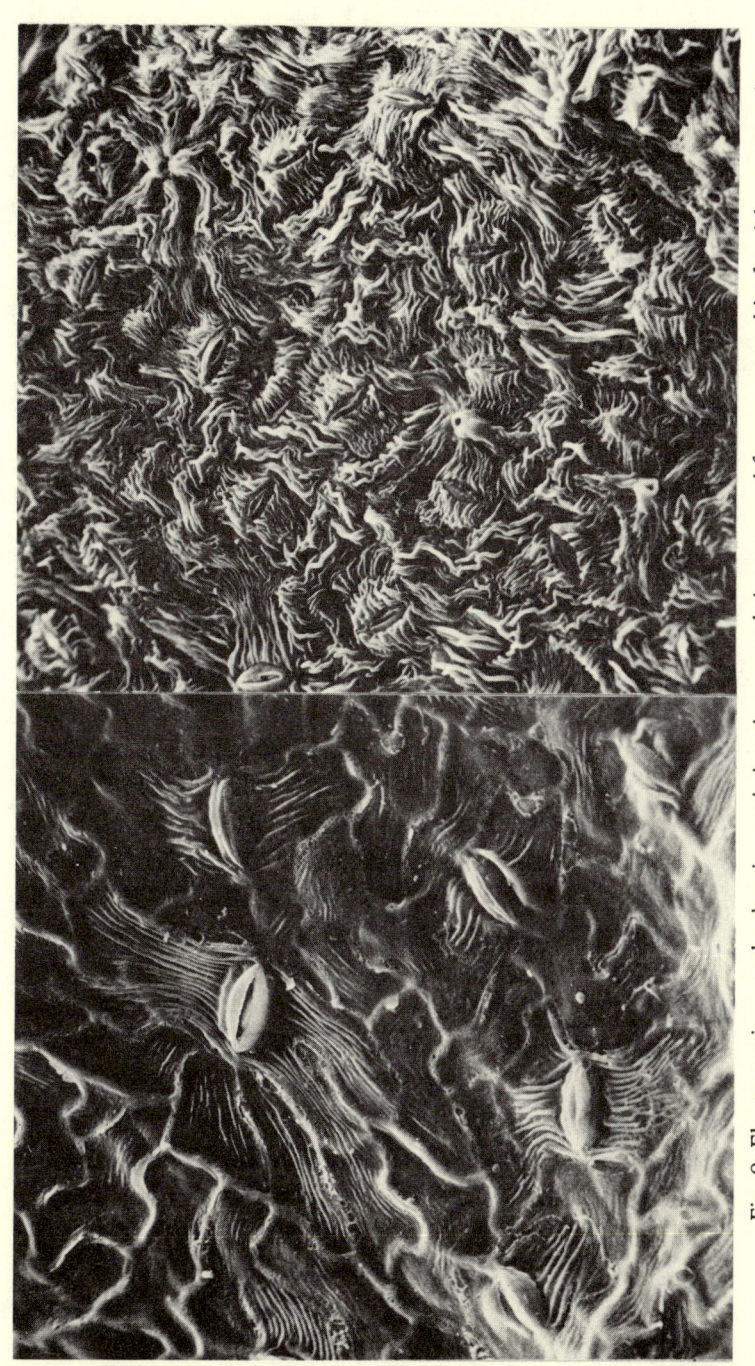

Fig. 8. Electron micrographs showing variations in stomatal size, stomatal frequency, and leaf cuticle structure in two hybrid *Populus* clones: *Populus deltoides* × *P. trichocarpa* (left); *P. alba* × *P. grandidentata* (right) (500x). *Photograph by S. Pallardy*

mesic. These included *Quercus macrocarpa, Q. velutina, Q. alba, Q. rubra,* and *Acer saccharum* (Wuenscher and Kozlowski 1971a). All species showed an increase in transpiration with increased leaf temperature. The rate of increase of transpiration resistance with rise in leaf temperature was greatest for *Q. velutina,* intermediate for *Q. macrocarpa* and *Q. rubra,* and least for *Q. alba* and *Acer saccharum.* Water use efficiency (the ratio of transpiration to photosynthesis) increased with leaf temperature up to 35°C and decreased at higher temperatures. Water use efficiency was greatest in *Quercus velutina,* intermediate in *Q. macrocarpa, Q. rubra,* and *Q. alba,* and least in *Acer saccharum.* At 40°C, water use efficiency of *Q. velutina* declined only slightly, but in the other four species it decreased greatly. These experiments showed that stomatal resistance decreased along the gradient from xerophytic to mesophytic species. In particular it emphasized that *Q. velutina* had the highest water use efficiency and was the most drought-resistant species studied. It fixed CO_2 rapidly while losing little water and therefore was successful on hot and dry sites (Wuenscher and Kozlowski 1971b).

Leaf Waxes. After stomata close, cuticular control of water is an important factor in drought avoidance. For example, transpiration rates of drought-evading plants with closed stomata often vary from 2 to 20 percent of the rates when stomata are open. On the other hand, mesophytes with closed stomata generally lose from 20 to 50 percent as much water as they do with open stomata, emphasizing the importance of leaf waxes in water retention.

Deposition of leaf waxes varies widely in different species and genetic materials. *Fraxinus americana* leaves had cuticular ledges overlapping large open stomatal pores, and they had thin leaf waxes. *Acer saccharum* leaves not only had a thick deposition of wax, but many stomatal pores were occluded with wax (Kozlowski, Davies, and Carlson 1974; Davies and Kozlowski 1974).

Gymnosperms also vary greatly in amount of leaf waxes. Needles of *Pinus sylvestris* had more wax than those of *P. radiata* (Leyton and Juniper 1963; Leyton and Armitage 1968).

Pine needles have a stomatal antechamber that increases the length of the diffusion path for water and carbon dioxide. These antechambers often are occluded with wax (Davies et al., 1974), thus increasing the tortuosity of the pathway and decreasing the cross-sectional area available for diffusion. The antechamber waxes often are more important than leaf-surface waxes in regulating transpirational water loss. It has been estimated that wax in stomatal antechambers of *Picea sitchensis* reduces transpiration by about two-thirds when the stomata are open (Jeffree, Johnson, and Jarvis 1971).

In addition to environmental control of deposition of leaf waxes, genetic control also occurs, with "normal" or glaucous forms being dominant and "glossy" or glabrous forms recessive. The differences in leaf appearance are due to variations in amount and structure of epicuticular wax (Hall et al. 1965). In some species at least, adaptive changes in leaf waxes occur in response to selection by some ecological variable. In Tasmania, for example, clinal analysis of glaucousness in several *Eucalyptus* species indicated high selective coefficients for genes controlling this characteristic. Nonglaucous (green) phenotypes were found in sheltered habitats; glaucous phenotypes in more exposed environments (Barber and Jackson 1957).

Root Growth. Because water in soil that is not penetrated by roots is largely unavailable to plants, trees with deeply penetrating and branching root systems absorb water most efficiently and prevent or postpone drought injury (Kozlowski 1972a). There are many examples of the importance of a high root-shoot ratio reflecting high water-absorbing capacity and relatively low transpiration capacity to survival of trees under drought conditions. For example, on dry sites *Eucalyptus socialis* outcompeted *E. incrassata* partly because of the higher root-shoot ratios and slower growth of aerial parts, thus preventing desiccation (Parsons 1969). Woods (1959) attributed the high degree of success of *Pinus caribaea* trees on dry sites to initiation of root growth soon after transplanting.

WATER EXCESS

Often an unfavorable internal water balance of trees results when too much standing water is present. Basically the problem is one of poor soil aeration as a result of displacement of oxygen in the soil by water. Thus, flooding can be very harmful, except for a few species.

The problem of poor soil aeration is common and

Water Supply and Tree Growth

does not always result only from excess water. Two general situations exist in the field: (1) where soil moisture is excessively high and little or no soil air is present, and (2) where gas exchange between the soil and atmosphere is insufficient to replace the oxygen depletion and carbon dioxide accumulates from respiration of roots and soil organisms. Compaction and trampling of soil, sidewalks, pavements, dense growth of grass, etc., can all cause poor soil aeration.

The composition of the atmosphere above ground is considerably different from that below ground level. The above-ground atmosphere contains approximately 21 percent oxygen, 78 percent nitrogen, and only 0.03 percent carbon dioxide. Since basic responses of protoplasm are more or less identical, it would be expected that free oxygen from the air must get down into the soil for normal root respiration to take place. The amount of carbon dioxide normally increases with depth of soil. Conversely, oxygen content decreases with soil depth.

There also are marked seasonal differences in composition of soil air. Soil air is richest in carbon dioxide in summer, less in autumn, still less in spring, and least in winter. Carbon dioxide content in the soil is increased by root respiration as well as by organic matter, manure, soil fungi, molds, bacteria and algae, together with soil animals from amoebae to rodents.

Water has only about 6 cc. of oxygen per liter when saturated during the growing season. Therefore, the water content and air content in soils are in inverse relation. The greater the water content, the smaller the air content and amount of available oxygen. Dry soils contain large amounts of air, and wet soils small amounts regardless of fineness.

Soil aeration is greatly influenced by soil texture. For example, aeration is commonly a problem in clay soil because most of the pores are so small that water does not drain readily. Such a soil holds little air, and movement of air between the soil and atmosphere is slow. By contrast, aeration problems are uncommon in sandy soils which contain considerable air, drain rapidly, and permit free movement of gases. The air capacity of heavy soils can be increased by adding organic matter, cinders, or sand.

Differences in aeration problems in soils of different texture were described by Kramer (1950). He pointed out that the root system of a tree 30 to 35 feet in height might thoroughly occupy the soil in a mass about 3 feet deep by 25 feet in diameter or a volume of 1,500 cubic feet. Many roots extend out much farther, but these will be disregarded. Such a root system would use all the oxygen in approximately 2,000 cubic feet of air during a growing season. A loam soil of good structure might contain about 20 percent by volume of air, or 300 cubic feet. Since this is only one-sixth of the total requirement, the oxygen must be completely replenished at least six times during the summer. A heavy clay soil of poor structure would hold 10 percent or less of air, or 100 to 150 cubic feet of air. The oxygen content of this soil would have to be replaced ten to fifteen times during the summer, or there would be a deficiency of oxygen for plant growth.

Beneath the surface of the soil, oxygen is usually suboptimal in concentration. Roots usually do not begin to show definite injury until the oxygen content of the soil atmosphere drops as low as 10 percent. Ordinarily the oxygen content of the upper levels of drained soils lies somewhere between the lower critical value of 10 percent and the 21 percent characteristic of free air. A reduction of soil oxygen to about 3 percent practically stops root growth in most plants. Because the oxygen content of soil drops abruptly to about 1 percent just above the water table, the roots of most land plants are restricted to soil horizons above this level.

Responses of Trees to Flooding

The poor soil aeration associated with flooding induces a number of complex physiological changes in trees that adversely affect growth. Root respiration changes from an aerobic to an anaerobic type, at least in part, with a consequent accumulation of many toxic compounds including nitrites, reduced forms of iron and manganese, hydrogen sulfide, ethylene, ethanol, and organic acids. Absorption of minerals requires energy released in respiration. Hence, mineral uptake is reduced by lack of oxygen in poorly aerated soils. In addition poor soil aeration affects mineral nutrition indirectly by inhibiting activities of soil organisms involved in nitrification.

Reduced water absorption and stomatal closure are among the earliest responses of trees to flooding. Subsequent changes include inhibition of shoot and root growth, alterations in root and stem morphology, formation of callus and adventitious roots, leaf yellowing and mottling as well as death and shedding of leaves. When flooding is severe and prolonged, trees are killed.

Both the time of occurrence of plant responses to flooding and extent of injury vary widely among species. Most gymnosperms are injured more than most angiosperms. Closely related species often exhibit wide variations in flooding tolerance. For example, *Nyssa aquatica* tolerates flooding better than *Nyssa sylvatica* (Hall and Smith 1955). *Salix* spp., *Ulmus americana, Quercus palustris,* and *Populus* spp. are injured less than *Acer saccharum, Fagus grandifolia,* or *Liriodendron tulipifera.* Orchard trees vary greatly in response to flooding, with species varying in the following order of decreasing sensitivity: olive > almond = peach = apricot > cherry > plum = citrus > apple > pear > quince (Rowe and Beardsell 1973). As pointed out by Gill (1970) the specific rankings of flooding tolerance of species as determined by different investigators should not be taken too seriously because of differences in age of plants used and in experimental conditions. Adult vigorous trees are less sensitive to flooding than are seedlings or overmature trees.

Flooding injures trees much more during the growing season than during the dormant season. Free water in the root zone of pecan trees in September and October induced leaf scorching and premature defoliation. By comparison, flooding in April and May did not injure trees (Alben 1958). *Liriodendron tulipifera* seedlings were killed after about three days of flooding in May and June but were not injured by flooding during the dormant season (McAlpine 1961). Greater flooding injury during the growing season is associated with high oxygen requirements of growing roots.

Mechanisms of Flooding Injury. The physiological dysfunctions associated with flooding that lead to injury are complex. Apparently, early symptoms of flooding are caused by decreased absorption of water and minerals. Later responses, however, such as epinasty and formation of adventitious roots have been correlated with high auxin levels. Other hormones are also involved in response to inundation of soils. Waterlogging reduces gibberellin levels in roots; hence it decreases the amount available for shoots. Flooding also reduces activity and production of cytokinins in roots. Chlorosis of lower leaves of flooded plants has been attributed, at least in part, to decrease in cytokinin synthesis in roots (Burrows and Carr 1969).

Three types of evidence indicate that ethylene is particularly important in inducing injury in flooded trees: (1) ethylene produces responses in unflooded plants that are similar to those associated with flooding, such as leaf epinasty, stem thickening, and leaf senescence and abscission; (2) ethylene production is stimulated in flooded soils and plants; and (3) application to plants of ethylene-releasing chemicals induces leaf abscission.

Smith and Russell (1969) demonstrated much higher accumulation of ethylene in soil under anaerobic than under aerobic conditions. Microorganisms appear to be responsible for ethylene production in anaerobic soils. Flooding increased ethylene accumulation in softwood cuttings of several species of woody plants (Table 6). Plant responses following application of Ethrel (which was absorbed and decomposed to ethylene in plants) closely resembled those caused by flooding (Kawase 1974).

TABLE 6
EFFECT OF SUBMERSION ON ETHYLENE CONCENTRATION IN *MALUS ROBUSTA* AND *LIGUSTRUM OBTUSIFOLIUM*

Cutting and Treatment	Ethylene concentration (ppm)			
	Before Treatment	After Treatment	5%	1%
Malus robusta				
Control	0.07	0.32	0.13	0.23
Submerged	0.06	1.66		
Ligustrum obtusifolium				
Control	0.06	0.50	0.40	0.74
Submerged	0.06	1.20		

SOURCE: Reprinted, by permission, from Kawase 1972, modified from table 1.
NOTE: Cuttings were completely submerged in water for 20 hrs. and controls were steeped upright in water 2.5 cm deep.

Both anatomical and physiological adaptations are important in tolerance of flooding. Formation of adventitious roots appears to be one of the most important adaptations. In some flooded species the amount of anaerobic respiration occurring in roots is supplemented by transport of oxygen from the atmosphere to the roots; for example, in *Salix atrocinerea* and *Nyssa sylvatica* var. *biflora*.

In many studies of tolerance to flooding, species that form adventitious roots at or below the water level survive best. The adventitious roots of flooded plants comprise a supplementary absorbing system in the somewhat aerobic zone, while the original root system does not function normally because of low oxygen tension (Gill 1970).

Nyssa sylvatica var. *biflora* seedlings developed tolerance to flooding through sequential anatomical and physiological adaptations. An early response to flooding was formation of water roots. These appeared to be beneficial because they occurred in the upper flood water where the oxygen content was higher and toxic compounds were lower than in the soil. The water roots also increased the absorbing surface for water and minerals. Under flooding, the newly initiated roots oxidized the rhizosphere whereas unflooded roots did not. Oxygen entered the stem through lenticels and appeared to be transported through the root cortex. The combined adaptations of increased anaerobic respiration, oxidation of the rhizosphere, and tolerance of CO_2 by the new roots appeared to account for the tolerance of flooding (Hook et al. 1970, 1971, 1972).

The Environmental Impact on Seeds and Seedlings

IN THE NEXT FEW DECADES FOREST BIOLOGISTS will be confronted with a broad spectrum of problems relating to natural regeneration of forest stands and increased production of high-quality planting stock. This seems obvious from our projected timber needs and strong emphasis in the Seaton report (Seaton *et al.* 1973) on the urgency of regenerating forest stands.

Unfortunately, forest trees face their greatest mortality risk when they are in the ungerminated embryo stage of the seed or in the young seedling stage of development. Many seeds fail to germinate because of one or more kinds of dormancy and others because of unsuitable environmental conditions. But even if seeds manage to germinate, the young seedlings live precariously for some time because they function at threshold levels of physiological growth requirements. Hence natural regeneration of forests and production of forest nursery stock depend on environmental conditions that are suitable for maintaining the germinating seed and young seedling in a physiologically efficient state. Even temporary mild environmental stresses can induce physiological dysfunction at this critical time in the life of a forest tree and lead to seed or seedling mortality.

Physiology of Seed Germination

In addition to minerals, phosphorus compounds, nucleic acids, and hormonal growth regulators, seeds contain variable amounts of reserve foods (carbohydrates, fats, proteins), with either carbohydrates or fats predominating, depending on species. These foods are stored in cotyledons and/or tissue surrounding the embryo (endosperm in angiosperms; megagametophyte in gymnosperms). In early stages of germination reserve foods are digested to simpler compounds that are then transported to growing parts of the embryo and used in growth. After a young seedling depletes the reserve foods of the seed, its continued growth depends on carbohydrates synthesized by epigeous cotyledons or, when cotyledons stay below ground as in oaks and walnuts, on carbohydrates synthesized by the first true leaves.

Seed Dormancy

Fortunately by the time the seeds of most temperate zone trees are shed they are programmed to have some degree of embryo dormancy. Such seeds do not germinate rapidly even under ostensibly optimal environmental conditions. A prolonged chilling requirement that is needed to break embryo dormancy delays germination in the field until the dangers of frost injury to young seedlings are over in the spring. Establishment and survival of certain species are further protected because release of seed dormancy from a proportion of the seeds in a given seedlot may be attenuated for several years. On the other hand, the occurrence of embryo dormancy is a serious problem to nursery operators and plant propagators who wish to produce uniform populations of plantable seedlings rapidly.

Environmental Control of Seed Germination

Seed germination involves a series of complicated and sequential physiological processes including hydration of tissues, increase in activity of enzymes and respiration, digestion of stored foods and translocation to growing embryo tissues, increase in organelles involved in metabolism, increased cell number and size, and differentiation of cells and tissues. The supply of energy in the form of adenosine triphosphate (ATP) necessary to drive germination processes becomes inadequate under conditions of environmental stress. Once the processes of germination are set in motion a continuously favorable environment for maintaining them is needed. Metabolic failure anywhere along the path of necessary events in germination is likely to kill the embryo. Seed germination is often prevented by even temporary unsuitable conditions of water supply, temperature, oxygen, and in the case of some seeds, light.

Only small amounts of water are needed to initiate protoplasmic hydration in seeds, initiate metabolic processes, and soften seed coats. Thereafter even temporary droughts may be catastrophic. After germination, large amounts of water must be continuously available to the growing seedling.

Temperature fluctuations also regulate germination. Prolonged chilling is required to break embryo dormancy. Thereafter much higher temperatures are needed to set metabolic processes in motion. Optimal germination temperatures for nondormant seeds vary markedly for seeds of different species as well as seed sources of the same species. Seed germination usually is more successful with diurnal temperature fluctuations than with constant temperatures.

Germination is often impeded by oxygen deficiency. Oxygen plays an important role as the electron acceptor in respiration, and oxygen is necessary for stimulating respiration in the very early phases of germination. Seeds with coats that are relatively impermeable to oxygen have especially high oxygen requirements.

Seeds of some species have light intensity or day-length requirements. Seeds of many species germinate in the dark as well as in the light; others require low illuminance. Seeds of most species germinate best in light periods of eight to twelve hours. However, some such as those of *Pseudotsuga menziesii* actually germinate better under long days or continuous light than under short-day conditions.

Seedling establishment in natural seedbeds varies greatly because of differences in water, temperature, and oxygen supply. Mineral soil and decayed wood usually are good natural seedbeds. Litter and duff are less suitable because they warm slowly, impede root growth, prevent seeds from contacting soil, and shade small seedlings. Natural seedbeds that are ideal for germination are not necessarily best for early growth of seedlings (Winget and Kozlowski 1965).

COTYLEDON PHYSIOLOGY

Work in our laboratory has emphasized the importance of healthy, physiologically active cotyledons from shortly after germination to seedling development. It has also emphasized extreme sensitivity of epigeous cotyledons to environmental stress. It appears that even temporary adverse environmental situations, such as low light intensity, water stress, and very low or high temperature, that inhibit physiological activity of cotyledons decrease growth of seedlings and often kill them (Kozlowski 1976a).

The specific physiological role of cotyledons in seedling development varies considerably among species. The cotyledons of *Juglans* and *Quercus* remain below ground and serve primarily as storage organs. Epigeous cotyledons of *Pinus* and *Cornus* store little food but play an important photosynthetic role shortly after they emerge from the ground. Cotyledons of exalbuminous seeds (lacking endosperm) such as *Fagus* and *Robinia* have dual roles of storage and photosynthesis. Cotyledons of species with endosperm also absorb reserves from endosperm and transfer them to growing axes. The size of cotyledons varies among species. Usually cotyledons of exalbuminous species are larger than those of species with seeds having considerable endosperm tissue (table 7).

Gymnosperms

Seed germination in pines is followed in order by development of cotyledons, primary needles, and secondary needles. After germination the seedling is a system of organs competing for carbohydrates. Seedling development is a continuous process with the site of synthesis of carbohydrates shifting from cotyledons to primary needles to secondary needles. Development of primary needles depends on contributions from cotyledons, and development of secondary needles depends on contributions from primary needles.

Our experiments have shown a strong influence of environment during the cotyledon stage on formation and expansion of primary needles of *Pinus resinosa*. Initiation and expansion of the first few primary needles depended on reserves in the megagametophyte of the seed, but development of most of the primary needles depended on carbohydrates produced by the cotyledons (Sasaki and Kozlowski 1968a, b; 1969; 1970).

There is a close association between the rate of cotyledon photosynthesis and seedling development in *Pinus resinosa*. Partial reduction of cotyledon photosynthesis by herbicides was followed by proportional decrease in expansion of primary needles and reduction in dry weight increment of seedlings (Sasaki and Kozlowski, 1968a, b). A strong influence of shoot

TABLE 7
Dry Weights of Cotyledons and Seedling Axes

Species	Seedling Age (Days)	Cotyledon Dry Weight (mg)	Seedling Axis Dry Weight (mg)
		Endosperm Absent	
Acer negundo	0	9.5	—
	2	8.8	3.0
	6	7.7	8.2
	10	6.4	40.3
	20	6.3	137.7
	30	6.0	284.0
Robinia pseudoacacia	0	6.1	—
	2	5.2	1.7
	6	4.7	3.6
	10	4.4	10.7
	20	4.1	42.0
	30	3.9	111.3
		Endosperm Present	
Ailanthus altissima	0	3.6	—
	2	3.3	1.5
	6	3.5	3.7
	10	4.6	5.5
	20	6.3	27.7
	30	6.3	71.8
Fraxinus pennsylvanica	0	1.1	—
	2	1.7	1.8
	6	2.8	4.4
	10	5.8	7.9
	20	6.3	24.7
	30	6.6	118.7

SOURCE: After Marshall and Kozlowski 1976c, modified from table 1.

TABLE 8

EFFECT OF LIGHT INTENSITY ON INITIATION AND EXPANSION OF PRIMARY NEEDLES OF *PINUS RESINOSA*

Treatment	Average Number of Visible Primary Needles After:					
	32 Days	39 Days	46 Days	53 Days	60 Days	
Dark, 25 °C	0	0	0	0	0	
70 ft-c (1.6×10^3 ergs/cm^2/s), 23–25 °C	0	0	0	0	0	
600 ft-c (1.3×10^4 ergs/cm^2/s), 25 °C	3.9 ± 0.24	7.6 ± 0.24	12.3 ± 0.30	17.3 ± 0.39	21.7 ± 0.48	
1200 ft-c (2.6×10^4 ergs/cm^2/s), 25 °C	7.4 ± 0.14	11.8 ± 0.32	16.3 ± 0.38	21.8 ± 0.43	25.3 ± 0.52	

SOURCE: Reprinted, by permission, from Kozlowski and Borger 1971, table 3.
NOTE: Values shown, means ± s.e.m. Time intervals indicate number of days after sowing.

environment early in ontogeny was shown on initiation of all but a few early-formed primary needle primordia and on expansion of all primary needles, including those formed early. Low temperatures or low light intensities during the cotyledon stage of development prevented initiation of most of the normal complement of primary needles (tables 8 and 9). However, when seedlings were returned to favorable light and temperature conditions, following prolonged exposure to low temperature or low light intensity, primordia of primary needles formed readily and subsequently expanded.

Angiosperms

We have divided the cotyledon stage of development of woody angiosperms exhibiting epigeous germination into storage-, transitional-, photosynthetic-, and senescent stages (Marshall and Kozlowski 1974a, b; 1975; 1976a,b,c; 1977).

Storage Phase: This phase is initiated after embryos imbibe water and the respiration rate of cotyledons increases. In seeds of some species *(Acer* spp., *Populus deltoides, Robinia pseudoacacia,* and *Ulmus americana),* endosperm is absent or nearly so, and the embryonic cotyledons of these species serve primarily as storage organs. Within this group of species the cotyledons vary greatly in size, growth capacity, dry weight, and carbohydrate, protein, lipid, and mineral nutrient contents. In another group of species *(Ailanthus altissima, Betula alleghaniensis, Fraxinus pennsylvanica, Gleditsia triacanthos,* and *Platanus occidentalis)* the embryonic cotyledons are embedded in endosperm and function as transfer as well as storage organs. The capacity of cotyledons for storage varies from small, as in *Betula* and *Platanus,* to large, as in *Gleditsia.*

The kinds and amounts of materials stored in cotyledons vary among species. Whereas embryonic cotyledons of *Acer negundo* and *Robinia pseudoacacia* contain considerable stored carbohydrate, those of *Ailanthus altissima* and *Fraxinus pennsylvanica* store little. The embryonic cotyledons of *Acer negundo, Ailanthus altissima, Fraxinus pennsylvanica,* and *Robinia pseudoacacia* store large amounts of lipids.

Greater amounts of major and minor elements are present in the embryonic cotyledons of *Robinia pseudoacacia* and *Acer rubrum* than in the smaller cotyledons of *Fraxinus pennsylvanica.* During the storage phase mineral elements are translocated from *Robinia pseudoacacia* and *Acer rubrum* cotyledons to seedling axes. By comparison cotyledons of *Fraxinus pennsylvanica* import mineral elements during this phase.

The epigeous cotyledons of *Acer saccharinum* fix CO_2 very early in germination and supply carbohydrates for growth. In *Gleditsia triacanthos* most of the early photosynthate is lost in respiration. Embryonic cotyledons of *Robinia* do not fix carbon early in germination.

Transition Phase. In this phase the cotyledons change from storage or transfer organs to assimilatory organs. This stage is characterized by expansion of cotyledons,

TABLE 9
EFFECT OF TEMPERATURE ON DEVELOPMENT OF COTYLEDONS AND PRIMARY NEEDLES OF *PINUS RESINOSA*

	Temperature (°C)				
	10	15	20	25	30
Average number of cotyledons per plant	6.54±0.09	6.52±0.09	6.46±0.08	6.60±0.09	6.36±0.07
Average number of primary needles per plant	2.74±0.38	9.06±0.44	16.74±0.39	16.62±0.50	17.58±0.66
Average length of cotyledons per plant (mm)	16.14±0.35	16.76±0.44	18.22±0.41	18.08±0.43	17.46±0.49
Average length of primary needles per plant (mm)	2.57±0.41	7.34±0.45	21.10±0.77	23.46±0.96	21.02±0.98
Average dry weight of cotyledons per plant (mg)	4.11	4.00	4.07	2.62	2.76
Average dry weight of primary needles per plant (mg)	0.34	2.13	9.01	7.10	8.21
Average dry weight of cotyledons plus primary needles (mg)	4.45	6.13	13.08	9.72	10.97

SOURCE: Reprinted, by permission, from Kozlowski and Borger 1971, table 2.
NOTE: Values shown, means ± s.e.m. ($n = 50$). Seedlings were grown from seed for 32 days at 23–25 °C and a light intensity of 70 ft-c (1.6×10^3 ergs/cm^2/s); they were subsequently grown for 44 days at indicated temperatures and a light intensity of 1200 ft-c (2.6×10^4 ergs/cm^2/s). At 10 °C, 27 plants out of 50 had expanded primary needles; at 15 °C, 48 plants out of 50 had expanded primary needles.

vacuolation of epidermal and mesophyll cells, appearance of functional stomata, synthesis of chlorophyll, and increase in photosynthesis of cotyledons. Respiration rates in the dark of cotyledons are high for most species, but decrease as reserves are depleted and cotyledons become increasingly photosynthetic.

Photosynthetic Phase. This stage begins with initial net CO_2 uptake. This occurs in about four days after radicle emergence in *Acer rubrum, Robinia pseudoacacia,* and *Ulmus americana* and a few days later in *Fraxinus pennsylvanica* and *Ailanthus altissima*. Photosynthetic capacity per cotyledon varied among species in the following order: *Robinia pseudoacacia* > *Ailanthus altissima* > *Fraxinus pennsylvanica* > *Ulmus americana.* The rate of photosynthesis of *Robinia pseudoacacia* cotyledons decreased rapidly after twelve days, whereas *Fraxinus pennsylvanica* cotyledons exhibited net CO_2 uptake for twenty-six days after radicle emergence but these rates were affected by environmental conditions. The importance of cotyledon photosynthesis and cotyledon reserves for seedling growth is demonstrated by the effects of excising cotyledons or inhibiting photosynthesis with DCMU (3-[3,4-dichlorophenyl]-1, 1, dimethylurea) (fig. 9; table 10). Excision of cotyledons of very young seedlings did not kill seedlings, but such treatments during the first ten days after radicle emergence inhibited growth of roots, hypocotyls, and foliage. These experiments emphasize that seedling growth and chances for survival of seedlings with photosynthetic cotyledons will be enhanced if care is taken to avoid injury to cotyledons and if environmental conditions are favorable for photosynthesis during the cotyledon stage.

Fig. 9. Importance of cotyledons to growth of *Ailanthus altissima* seedlings. *Left to right:* 21-day-old seedlings that had cotyledons removed at 2, 4, 6, 8, or 10 days after radicle emergence. The intact control seedling is on the far right. Reprinted from Marshall and Kozlowski 1976a.

Senescent Stage. This stage is characterized by marginal yellowing and decreased break-strength of cotyledons. Carbohydrate content of cotyledons of *Acer negundo, Ailanthus altissima, Fraxinus pennsylvanica,* and *Robinia pseudoacacia* tends to increase during cotyledon senescence. Some mineral elements are translocated from senescing cotyledons.

Environmental Impacts on Seedlings

In addition to light, water, and temperature effects, a number of naturally occurring and applied chemical substances suppress seed germination and inhibit seedling growth. Such substances include some air pollutants, insecticides, fungicides, herbicides, and fertilizers, as well as various inhibitors in plants (Kozlowski 1971a).

Air Pollutants

Work in our laboratory has emphasized that *Pinus resinosa* seedlings in the cotyledon stage of development are very sensitive to SO_2 (Constantinidou, Kozlowski, and Jensen 1976). Two-week-old seedlings were exposed to four concentrations of SO_2 (0.5, 1, 3, or 4 ppm) at four exposure times (15, 30, 60, or 120 minutes). After fumigation, these seedlings, as well as unfumigated control seedlings, were kept in an environmentally controlled growth chamber for eleven weeks. At that time the cotyledons were senescing, and the plants were harvested.

Seedling development was inhibited by SO_2 at all concentrations for 120 minutes and at 1 ppm and above for 30 minutes or longer. The adverse effects were proportional to SO_2 concentration and duration of exposure. The fumigations induced chlorosis in both cotyledons and primary needles, slowed expansion of primary needles, inhibited dry weight increment of both cotyledons and primary needles, and induced necrosis of tips of cotyledons and primary needles.

Visible injury to both cotyledons and primary needles occurred only after fumigation with the higher concentrations of SO_2 and longer exposure times. However, physiological efficiency of cotyledons was impaired even when visible injury did not occur. This was shown by decreased chlorophyll content as well as decreased dry weight increment of both cotyledons and primary needles. Since some time lag between impairment of physiological efficiency and decreased development of primary needles might be expected, the early detection of such a response after fumigation with SO_2 for relatively short times is of particular interest. The data suggested that continuous point-source fumigation with SO_2 at much lower dosages than were used in our experiments would have important inhibitory effects on seedling development and regeneration of pine communities.

Seedlings well beyond the cotyledon stage are also very susceptible to pollution. We found that four-month-old *Ulmus americana* plants were very sensitive to air pollutants. Sulfur dioxide (2 ppm for 6 hr), ozone (0.9 ppm for 5 hr), and sulfur dioxide-ozone (2 ppm SO_2 and 0.9 ppm O_3 for 5 hr followed by 1 hr exposure to 2 ppm SO_2) significantly reduced growth of shoots and roots, total nonstructural carbohydrates, and protein contents. Changes in lipid contents were small and not significant. The SO_2-O_3 mixture had the greatest effect on metabolites and growth, followed by SO_2 and O_3, in that order. Reductions in carbohydrate

TABLE 10
Dry Weights (mg) of Various Plant Parts from *Robinia pseudoacacia* Seedlings with Cotyledons Excised or Treated with DCMU at Different Seedling Ages

Plant Part	Seedling Age Days	Seedling Age at Decotylization, Days				Seedling Age when DCMU Applied, Days				
		2	6	10	Control	4	6	8	10	Control
Roots	14	2.1	3.2	7.1	22.7	1.6	1.7	2.9	17.2	29.4
	21	5.6	7.8	14.2	45.4	3.2	17.5	8.2	28.1	45.6
	28	18.5	12.1	27.7	59.5	4.0	10.5	15.2	26.6	45.7
Hypocotyls	14	1.0	1.7	1.9	3.0	1.9	1.5	1.9	2.7	3.5
	21	1.8	2.1	2.9	4.9	2.2	2.6	2.4	3.9	5.5
	28	2.6	2.3	3.2	7.6	1.6	3.0	3.3	3.1	7.0
Foliage	14	2.6	4.2	4.4	11.4	1.9	1.5	3.0	11.7	11.5
	21	8.0	9.7	14.0	22.6	11.1	9.9	13.0	17.4	22.6
	28	16.0	14.6	23.7	71.2	6.9	20.1	26.1	21.8	51.3

SOURCE: After Marshall and Kozlowski 1976a, table 1.

and protein contents of leaves and roots were found within a day after fumigation, inhibition of leaf expansion within one week, and reduction in dry weight increment of stems and roots within five weeks.

Herbicides

Some weed-controlling chemicals have been found very useful in production of nursery stock. Others pose problems because of their harmful effects on physiological processes in young trees. Ideally, for new seedbeds, nursery managers prefer selective preemergence herbicides that can be applied immediately after seeding to control weeds for at least one season.

Susceptibility of tree seedlings to herbicides, however, varies greatly with the chemical used; the rate, method, time, and number of applications; the species and even variety of trees; age of trees; soil type; weather; and other factors. Because seedlings are most susceptible when in the cotyledon stage, certain herbicides that can be used in transplant beds or in plantation establishment cannot be used in seedbeds (Kozlowski and Sasaki 1970).

Another problem is the possible persistence of herbicides and accumulation of toxic residues in nurseries where beds are used repeatedly to grow a variety of tree species for different lengths of time. Longevity of different herbicides in soil varies greatly and is influenced by leaching, microbial action, absorption, volatilization, and chemical reaction.

Some herbicides are very toxic to young tree seedlings, while others are not. We found, in particular, that Dacthal (DCPA) (see App. 1 for composition of herbicides mentioned in the text) applied at rates up to four pounds per acre as a preemergence and postemergence spray to *Pinus resinosa* seedbeds did not affect seedling emergence and did not induce phytotoxic symptoms in young pine seedlings.

After 2½ months, the treated seedlings were comparable to untreated control seedlings in stand density, size, color and general appearance. Furthermore, Dacthal effectively controlled several weeds common to forest nurseries, including carpetweed *(Mollugo verticillata)*, purslane *(Portulaca oleracea)*, chickweed *(Stellaria media)*, witchgrass *(Panicum capillare)*, and others.

By comparison, a number of herbicides were very toxic, as a result of direct suppression of seed germination, subsequent toxicity to young seedlings, or both. Although it often is assumed that the toxic agent of commercial herbicide formulations is entirely in the "active" ingredients, toxic effects sometimes also are traceable to so-called inert ingredients and to interactions of active and inert ingredients (Sasaki and Kozlowski 1968c,d).

Some herbicides greatly suppress seed germination and others have no serious effects but may injure recently emerged seedlings. For example, seed germination was influenced only very slightly by soil-incorporated DCPA, CDEC, EPTC, and several triazine herbicides (Simazine, Atrazine, Prometryne, and Propazine). In contrast all these herbicides except DCPA caused seedling mortality and decreased dry weight increment of seedlings in varying amounts (tables 11–14). EPTC and CDAA at high dosages were very toxic to young seedlings, whereas DCPA showed no deleterious effects at any dosage. Toxicity of soil-incorporated herbicides was generally much greater than when the same herbicides were applied to the soil surfaces. Toxicity of triazine herbicides varied in the following order of decreasing toxicity: Atrazine, Simazine, Prometryne, Propazine, and Ipazine. Often toxicity was greatly delayed. For example, more seedlings died in the last twenty days of the experiment than in the first ninety days, emphasizing the danger of evaluating herbicide toxicity in short-time experiments only (Kozlowski and Torrie 1965).

Seedlings emerged uniformly from soil with incorporated triazine herbicides, but shortly the cotyledons became twisted and incurled. Eventually many of them died. It should be added, however, that Simazine applied to the soil surface can be used very successfully to control weeds around older pines after they are outplanted. The absorbing roots of such trees generally are below the zone of soil containing the herbicide and therefore escape injury.

Several herbicides tested in our laboratory caused marked morphogenic changes during seedling development. Plants whose seeds had been treated with CDEC or EPTC had fused cotyledons and those treated with 2,4-D had swollen stems and chlorotic, shriveled cotyledons.

At concentrations of 25 ppm or higher 2,4,5-T maintained in direct contact with seeds and recently germinated seedlings of *Pinus resinosa* caused abnor-

TABLE 11
Effect of Soil Incorporation of Triazine Herbicides at Different Dosages on Germination of *Pinus resinosa* Seeds

Treatment	Germination after Planting*		
	33 days	43 days	73 days
(lb./acre)		%	
Control†	22.5	47.0	72.6
Ipazine			
2	38.0	62.8	72.1
4	17.5	46.4	65.6
8	34.3	52.1	68.0
16	16.9	56.1	73.9
Propazine			
2	39.0	71.8	82.9
4	42.8	65.7	79.5
8	29.7	56.9	70.6
16	34.0	60.4	77.9
Prometryne			
2	17.4	41.8	69.1
4	23.6	55.3	70.6
8	18.1	43.9	65.2
16	15.6	52.3	71.3
Simazine			
2	35.8	58.0	72.5
4	9.5	32.8	62.4
8	27.3	58.2	74.7
16	18.7	45.8	72.4
Atrazine			
2	32.6	62.6	77.0
4	29.9	53.6	77.5
8	21.2	54.7	75.1
16	37.0	69.3	81.0
Coefficient of variation	64%	32%	19%

Source: Reprinted, by permission, from Kozlowski and Torrie, © 1965 The Williams & Wilkins Co., Baltimore, Md., table 3.
*Values are averages of 10 replicates of 100 seeds each.
†No herbicide.

TABLE 12
Effect of Soil Incorporation of DCPA, CDEC, EPTC, and CDAA at Different Dosages on Germination of *Pinus resinosa* Seeds

Treatment	Germination after Planting*		
	23 days	35 days	54 days
(lb./acre)		%	
Control†	27.8	39.4	49.4
DCPA			
2	34.1	59.2	69.4
4	30.0	53.7	66.4
8	32.6	48.0	60.0
16	32.9	48.1	60.8
CDEC			
2	34.8	51.6	63.5
4	20.8	36.6	46.6
8	30.5	56.4	63.1
16	24.0	37.7	48.9
EPTC			
2	23.5	47.1	59.2
4	32.3	51.3	63.2
8	38.2	57.7	62.8
16	27.6	46.1	53.6
CDAA			
2	41.7	55.5	66.6
4	27.4	45.0	51.0
8	26.7	56.0	63.9
16	29.5	55.6	61.9
Coefficient of variation	69%	69%	40%

SOURCE: Reprinted, by permission, from Kozlowski and Torrie, © 1965 The Williams & Wilkins Co., Baltimore, Md., table 2.
*Values are averages of 10 replicates of 100 seeds each.
†No herbicide.

mal development of seedlings. Responses to 2,4,5-T included inhibition of root elongation; proliferation, expansion, and collapse of parenchyma cells in the stem, root, and cotyledons; formation of callus tissue; and inhibition of formation of primary needle primordia and their expansion; as well as distortion of primary needles and fusion of primary needles to cotyledons (Wu and Kozlowski 1972).

The preoccupation with sensitivity of pine seedlings in the cotyledon stage to herbicides should not detract from the fact that older plants are also sensitive, although less so, to certain herbicides. For example, when we applied triazine herbicides to *Pinus resinosa* seedlings and transplants in standard nursery beds (surface application in June up to 4 lbs/acre on Plainfield sand) we found considerable injury and growth inhibition in 1-0 plants, less in 2-0 plants, and none in transplants (Winget, Kozlowski, and Kuntz 1963). Decreased growth of three-year-old *Pinus resinosa* plants following spray application or soil incorporation of Monuron, Atrazine, or EPTC was associated with inhibition of photosynthesis.

Herbicide Selectivity. It cannot be emphasized too strongly that observed toxicity of herbicides to trees is influenced by a host of herbicide, plant, and environmental factors. Toxicity or selectivity of a herbicide should not be considered as a generalized have or have-not attribute. We often are asked whether a particular herbicide is toxic to a given species. Such a question is not easily answered unless we also know the soil type, the environment, the herbicide dosage,

TABLE 13
Effect of Soil Incorporation of Herbicides on Survival and Dry-weight Production of *Pinus resinosa* Seedlings

Treatment (lb./acre)	Dry Weight† g.	As % of control
Control‡	13.43	100
DCPA		
2	11.19	83.3
4	16.15*	120.3
8	13.45	100.0
16	11.94	88.9
CDEC		
2	9.04**	67.3
4	10.36**	77.1
8	11.21	83.5
16	9.90**	73.7
EPTC		
2	10.89**	81.1
4	7.48**	55.7
8	3.64**	27.1
16	3.33**	24.8
CDAA		
2	6.64**	49.4
4	5.87**	43.7
8	5.54**	41.3
16	1.98**	14.7

SOURCE: Reprinted, by permission, from Kozlowski and Torrie, © 1965 The Williams & Wilkins Co., Baltimore, Md., table 4.
†Of living and dead seedlings in 10 cartons. Data taken 54 days after planting. ** = significantly different from control at the 1% level, and * = at the 5% level.
‡No herbicides.

the growth stage of the treated plants, the manner of herbicide application, etc. All of these will affect the plant response. Furthermore, if we are dealing with herbicides that persist in the soil we need to know the probable amplitude of environmental factors for a long time after the herbicides are applied.

There are several types of herbicide selectivity. Most important among these are chemical selectivity, physical selectivity, physiological selectivity, and placement selectivity.

Chemical Selectivity. True chemical selectivity is thought to derive from differences between enzyme systems of different species. Substance X may have no effect on species A and yet be extremely toxic to species B. A wide variety of biochemical processes, including hundreds of reactions that are catalyzed by enzymes, take place by virtue of the matching in size and shape of a cavity in a host molecule with the form of a guest molecule, the combination forming what is known as an inclusion compound. Slight differences in shape or size of one or more compounds of different enzyme systems may determine whether or not a reaction takes place with the molecules of a given herbicide. When the tolerance shown by a plant to low levels of an herbicide gives way to susceptibility at higher levels, it may be assumed that the herbicide inactivation capacity of the plant enzyme system has been exceeded.

Physical Selectivity. Selectivity sometimes depends on differences between the morphology, nature of leaf and bark surfaces, or life form of crop and weed spe-

TABLE 14
EFFECT OF SOIL INCORPORATION OF TRIAZINE HERBICIDES ON SURVIVAL AND DRY-WEIGHT PRODUCTION OF *PINUS RESINOSA* SEEDLINGS

Treatment	Dry Weight†		
	Shoots of live seedlings‡	As % of	Per live seedling
(lb./acre)	(g.)	control	(g.)
Control§	4.046	100.0	0.0117
Ipazine			
2	4.048	100.0	0.0116
4	5.594	138.3	0.0106
8	3.059	75.6	0.0080
16	0.875**	21.6	0.0064
Propazine			
2	3.277	81.0	0.0087
4	1.284**	31.7	0.0068
8	0.516**	12.8	0.0058
16	0.080**	0.2	0.0057
Prometryne			
2	1.206**	29.8	0.0086
4	0.724**	17.9	0.0066
8	0**	0	0
16	0**	0	0
Simazine			
2	0.497**	12.2	0.0057
4	0.342**	8.5	0.0054
8	0.213**	5.3	0.0056
16	0**	0	0
Atrazine			
2	0.010**	0	0.0051
4	0.320**	7.9	0.0052
8	0.068**	0.2	0.0056
16	0.076**	0.2	0.0051

SOURCE: Reprinted, with permission, from Kozlowski and Torrie, © 1965 The Williams & Wilkins Co., Baltimore, Md., table 6.
†Data taken 110 days after seeds were planted.
** = significantly different from control at the 1% level, and * at the 5% level.
‡In 10 cartons.
§No herbicide.

cies. Leaves of some species shed herbicide sprays whereas others retain large amounts. Although the herbicide might be equally toxic to several species given equal ease of entry, a form of selectivity operates to give differential killing. Much of the greater susceptibility of angiosperms than of hardened-off gymnosperms to hormone-type herbicides (2,4-D) is certainly a consequence of the broader and more hairy leaves of hardwoods in contrast with the thick-cuticled leaves of conifers. The marked differences between gymnosperms with rapidly growing shoots and hardened-off gymnosperms in their susceptibility to hormone type herbicides is also largely an expression of physical selectivity. In the western states, a margin of selectivity exists between *Pinus ponderosa* and dwarf mistletoe because the mistletoe has a cuticle that is penetrated more easily than that of *Pinus ponderosa* by certain sprays.

Physiological Selectivity. This type depends on the growth stage of the plant. For example, when gymnosperm shoots are expanding, they are very susceptible to 2,4-D. A good example of physiological selectivity is the response of *Abies grandis* to 2,4,5-T. When applied during the dormant season injury does not occur, whereas applications during the growing season readily kill trees.

If the phenology of species is such that a herbicide treatment is applied when one species is relatively immune while another is relatively susceptible, then selectivity will operate to the detriment of the latter. Amino-triazole can be used to release shielded *Picea glauca* trees from grass and hardwood competition from midsummer onward, but earlier in the season the herbicide is so toxic to *Picea* that even careful shielding cannot prevent damage in field operations.

Placement Selectivity (Selective Application). Any herbicide, even a nonselective one, can be used selectively by applying it carefully to weed species while leaving crop trees untreated. For example, *Pinus strobus* has been released from hardwoods by applying Fenuron around each weed stem. Fenuron has also been carefully placed around *Picea* trees at the time of planting to reduce shrub competition.

It should be remembered that in practice many herbicides are toxic to different degrees. They are selective only in the sense that more herbicide is needed to control one plant species than another.

Allelopathy

Seed germination and seedling growth are inhibited not only by applied chemicals but also by a variety of naturally occurring compounds that are released to the soil from roots and aerial tissues of neighboring plants (Rice 1974). In 1832 DeCandolle called attention to injury to plants by root excretions of other plants. Since that time there have been periodic surges of interest in the deleterious biochemical effects of one plant on another. Unfortunately many of the early experiments tended to overemphasize the ecological significance of toxins released by plants on the basis of laboratory experiments. Lerner and Evanari (1961), for example, found in laboratory experiments that leaves of *Eucalyptus rostrata* contained substances that inhibited seed germination. However, tests of soil from beneath *Eucalyptus* trees showed that these allelopathic chemicals did not accumulate to toxic levels. In the last two decades there has been a renewed interest in allelopathy, especially in forest ecosystems, and enough information has now accumulated to indicate that allelopathic effects play an important role in establishment of some forest stands.

Allelopathic compounds are of widespread occurrence in both herbaceous and woody plants. Their effects usually are exerted on other species but autotoxicity also occurs, as in *Eucalyptus*. In general allelopathic compounds occur in plants in ways that protect the plant against their effects. Many allelochems occur as glycosides. The toxic substance, by combining with a sugar, for example, may become innocuous within the plant (Whittaker and Feeny 1971).

Allelopathic compounds may be released to the soil by leaching, volitization, excretion, exudation, and decay, either directly or by activity of microorganisms. Among the naturally occurring compounds which appear to have inhibiting effects on seed germination and growth of neighboring plants are phenolic acids, coumarins and quinones, terpenes, essential oils, alkaloids, and organic cyanides. Allelopathic chemicals are ecologically important because they influence succession, dominance, vegetation dynamics, species diversity, structure of plant communites, and productivity (Whittaker 1970).

Many studies have shown that establishment and

growth of forest trees are inhibited by extracts of herbaceous or shrubby woody plants. Fisher (1976) cited many examples of inhibitory allelopathic effects of a variety of species on establishment and growth of forest trees. These included effects of *Kalmia angustifolia* on *Picea mariana;* *Calluna vulgaris* on *Picea abies;* *Prunus pumila, P. serotina, Salix pellita, Solidago uliginosa,* and *Solidago juncea* on *Pinus banksiana;* *Aster* and *Solidago* spp. on *Liriodendron tulipifera;* *Festuca arizonica* and *Muhlenbergia montana* on *Pinus ponderosa;* *Aster novaangliae, Solidago canadensis, Solidago graminifolia,* and *Hieracium pratense* on *Acer saccharum.* Of particular interest to the western United States are studies showing that allelochems from dead fronds of bracken fern, *Pteridium aquilinum,* a second stage successional species, impede establishment of *Pseudotsuga menziesii* (Stewart 1975). Radicle elongation of *Pseudotsuga menziesii* is also inhibited by leaf and litter extracts from *Arbutus menziesii, Acer circinatum, Sambucus racemosa,* and *Symphoricarpos albus* and by litter extracts, but not leaf extracts, of *Cornus nuttallii, Abies procera, Physocarpus capitatus,* and *Rubus parviflorus* (del Moral and Cates 1971).

More and more evidence is accumulating that interference from field plants is a deterrent to invasion by trees. Adverse climatic and competitive forces, together with allelopathic effects of old field weeds and their decomposing organic material, also appear to be very important. This was emphasized by Horsley's (1977) study of allelopathic inhibition of *Prunus serotina* seedlings in the Allegheny Plateau in northwestern Pennsylvania. In that region virgin *Tsuga canadensis-Fagus grandifolia* and *Fagus grandifolia-Acer saccharum* forests were almost completely clearcut between the 1880s and early 1930s. On most areas a secondary succession of *Prunus serotina, Acer rubrum, Acer sacharum, Fraxinus americana,* and associated species developed, and these species are present today. However, forests failed to return to areas on which fire after logging destroyed most of the seedlings and much of the organic mantle. Extensive areas are now without trees or have only a few scattered *Prunus serotina* and *Acer rubrum* trees. Several possible reasons have been advanced for the lack of regeneration: lack of seed source, lack of advance reproduction, browsing by animals, high soil temperatures, frost, low soil moisture supplies, shade, and lack of nutrients. Although these factors apparently contribute to the problem, they do not appear to be the major causes of failure of regeneration. Large numbers of *Prunus serotina* seedlings are present under isolated trees. They grow slowly and do not become established.

In a series of well-designed experiments Horsley (1977) tested the effects of allelochems, produced by several herbaceous ground cover plants, on germination of *Prunus serotina* seeds, growth of young seedlings with cotyledon reserves, and growth of seedlings after cotyledon reserves were exhausted.

The effects of foliage extracts and root washings of bracken fern *(Pteridium aquilinum),* wild oat grass *(Danthonia compressa),* goldenrod *(Solidago rugosa),* and flat topped aster *(Aster umbellatus)* were studied separately.

Foliage extracts of fern, goldenrod, and aster variously inhibited seed germination (table 15). Only aster foliage extract inhibited growth when cotyledons contained reserves (tables 16 and 17). The failure of *Prunus serotina* seedlings to grow was attributed mainly to effects of allelochems after seedlings exhausted cotyledon reserves. At that stage of seedling development, foliage extracts of fern, goldenrod, grass, or aster had inhibitory effects on growth (table 18). Also root washings of goldenrod and aster significantly reduced growth of *Prunus serotina* seedlings, whereas single clumps of grass or fern did not (table 19).

The allelochems apparently are bound to the soil complex and are not leachable. The fact that *Prunus serotina* seedlings survive and grow after the herbaceous vegetation is removed, but not until the second year after removal, indicates that allelopathic residues are present in the soil. During the one-year lag the toxic residues apparently are neutralized, destroyed, or leached away to eventually permit regeneration of *Prunus serotina*. If the herbaceous plants are prevented from recolonizing these sites, growth of *Prunus serotina* probably will return to normal rates.

The effects of allelopathic chemicals on neighboring plants are modified by several factors such as soil moisture, soil type, soil microflora, etc. Terpenes produced by *Eucalyptus camuldulensis* influenced the annual grassland flora only after becoming adsorbed to soil parti-

TABLE 15
GERMINATION OF *PRUNUS SEROTINA* SEED STRATIFIED IN PEAT MOISTENED WITH 5, 25, 50, or 100% FOLIAGE EXTRACT OF FERN, GRASS, GOLDENROD, OR ASTER AS A PERCENTAGE OF GERMINATION OF SEED STRATIFIED IN PEAT MOISTENED WITH DISTILLED WATER

Extract	Concentration, %	Germination, %
Fern	5	46±5**
	25	48±19**
	50	35±13**
	100	41±15**
Grass	5	70±22*
	25	93±16
	50	92±21
	100	119±32
Goldenrod	5	73±22*
	25	68±7**
	50	132±12**
	100	57±21**
Aster	5	54±19**
	25	50±12**
	50	73±18*
	100	37±8**

SOURCE: Reprinted, by permission, from Horsley 1977, table 2.
NOTE: Each figure is the mean of five replicates ± 1 SD.
*Significant at the 0.05 level of probability.
**Significant at the 0.01 level of probability.

cles, whereas phenolic acids influenced growth more directly. Well-drained light soils did not concentrate these toxins and favorable aeration permitted their rapid degradation. In dry soils competition for water combined with allelopathic effects of terpenes and phenolics to produce extensive bare zones. In wet soils some growth of annual herbs occurred despite the presence of toxins. Readily available soil water apparently lessened the possibility that inhibited plants would be killed by drought. Heavy rains also favored degradation of allelopathic chemicals and washed them deep into the soil and diluted them (del Moral and Muller 1970).

More research is needed on the implication of allelopathy in forest ecosystems. We need better data on the influence of environmental factors on allelopathy. Greater understanding of precise mechanisms of allelopathic interactions may influence the development of methods for old field planting. Both the capacity to produce allelochems and the degree of susceptibility

TABLE 16

MEAN SHOOT HEIGHT AND NUMBER OF FIRST-ORDER LATERAL ROOTS OF *PRUNUS SEROTINA* SEEDLINGS GROWN ON COTYLEDONARY RESERVES FOR 17 DAYS IN SAND WATERED WITH 100% FOLIAGE EXTRACT OF FERN, GRASS, GOLDENROD, OR ASTER OR DISTILLED WATER

Extract	Shoot Height, mm	No. First-order Lateral Roots
Distilled-water control	49±7	20±5
Fern	52±4	22±8
Grass	53±5	21±4
Goldenrod	52±4	21±5
Aster	32±4**	10±2**

SOURCE: Reprinted, by permission, from Horsley 1977, table 3.
**Significantly different from control at the 0.01 level of probability.

to them are genetically controlled. Information on the nature of allelopathy would help tree breeders decrease or increase allelopathic effects as desired for regenerating forest stands.

Physiological Characteristics of Planting Stock

Forest biologists can help to identify attainable features of nursery-grown and containerized seedlings that will improve their capacity to cope with environmental stresses after outplanting. Of primary concern is the capacity of planting stock to withstand drought and low temperature.

Seed Source

Natural selection has tended to produce natural populations of trees that have become adapted to the conditions in which they have evolved. Recognizing this, forest biologists can do much to avoid harmful effects of environmental stresses by giving careful attention to sources of seed for planting on specific sites, especially periodically dry or cold ones. Unless better sources of seed are known, attention should be given to using seed from trees of native stock growing as near as possible to the site to be planted. The U.S. Forest Service suggests that, if possible, planting stock should be produced from seed collected within 100 miles horizontally and 1000 feet vertically of the place where it will be used.

For species with extensive north–south ranges the

TABLE 17
HEIGHT (MM) OF *PRUNUS SEROTINA* SEEDLINGS GROWN ON COTYLEDONARY RESERVES FOR 14 DAYS ON THREE SOILS WATERED WITH 100% FOLIAGE EXTRACTS OF FERN, GRASS, GOLDENROD, OR ASTER OR DISTILLED WATER

Extract	Sandy Loam Soil, Orchard Stand	Sandy Loam Soil, Allegheny Hardwood Stand	Silt Loam Soil Agricultural Quality
Distilled-water control	24±10	28±6	25±12
Fern	31±12*	31±7	32±18
Grass	28±10	29±11	37±3
Goldenrod	32±3*	39±13	34±9
Aster	17±8	9±4**	34±14

SOURCE: Reprinted, by permission, from Horsley 1977, table 4.
*Significantly different from control at the 0.05 level of probability.
**Significantly different from control at the 0.01 level of probability.

southern seed sources tend to grow faster, leaf out later in the spring, continue shoot growth later into the autumn, and are less cold resistant. As emphasized by Wright (1976) these trends largely reflect adaptations to cold and warm regimes. Similar trends are evident in the Western states in proceeding from the relatively cold interior to the relatively warm Pacific Coast. Other trends are evident in progressing from a dry to a wet region. In comparison with trees from a dry region, those from a wet one generally grow faster and are less deeply rooted.

Elevational trends also are important, particularly in the Western states. A 3,000-foot elevational difference often reflects a climatic difference similar to that occur-

TABLE 18
GROWTH OF *PRUNUS SEROTINA* SEEDLINGS WATERED WITH FERN, GRASS, GOLDENROD, OR ASTER FOLIAGE EXTRACTS IN COMPLETE NUTRIENT SOLUTION, AS A PERCENTAGE OF CONTROL SEEDLINGS WATERED WITH COMPLETE NUTRIENT SOLUTION

Measurement	Control (basis)	Fern			Grass			Goldenrod			Aster		
		5	25	50	5	25	50	5	25	50	5	25	50
Total height	100	55*	—	31*	—	65*	60*	56**	53**	47**	74	46*	37**
No. nodes	100	79	—	68*	—	83	83	71*	83	67*	83	75*	62**
Internode length	100	68	—	52	—	77	75	80	65	70	92	73	57
Shoot dry weight	100	52*	—	26**	—	68	56*	53*	45*	34**	68	52*	28*
Root dry weight	100	81	—	69	—	119	100	101	98	86	122	128	77
Shoot plus root dry weight	100	66	—	53*	—	100	84	83	78	67	101	99	46*

SOURCE: Reprinted, by permission, from Horsley 1977, table 5.
NOTE: Seedlings had exhausted cotyledonary reserves before treatment.
*Significant at the 0.05 level of probability.
**Significant at the 0.01 level of probability.

TABLE 19

GROWTH OF *PRUNUS SEROTINA* SEEDLINGS RECEIVING ROOT WASHINGS OF FERN, GRASS, GOLDENROD, OR ASTER IN COMPLETE NUTRIENT SOLUTION, AS A PERCENTAGE OF CONTROL SEEDLINGS RECEIVING ROOT WASHINGS OF *PRUNUS SEROTINA* IN COMPLETE NUTRIENT SOLUTION

Measurement	Control (basis)	Fern	Grass	Goldenrod	Aster
Total height	100	92	99	53**	48**
No. nodes	100	84	97	69**	69**
Internode elongation	100	106	104	74**	70**
Shoot dry weight	100	88	101	79	26**
Root dry weight	100	95	86	57*	23**
Shoot plus root dry weight	100	90	93	68	24**

SOURCE: Reprinted, by permission, from Horsley 1977, table 6.
NOTE: Seedlings had exhausted cotyledonary reserves before treatment. Each figure is the mean of four plants.
*Significant at the 0.05 level of probability.
**Significant at the 0.01 level of probability.

ring in a few hundred miles of level ground. Selection may therefore operate to induce significant differences, in responses to stress, in trees of high and low elevations. However, the rate of gene exchange is much higher between different elevations on the same mountain than between populations a few hundred miles apart (Wright 1976). Such gene exchange tends to keep the high and low elevation populations of trees from diverging genetically. Hence, elevational clines in the Western states are well known (Squillace and Silen 1962).

As pointed out in the first lecture, trees from southern seed sources are more susceptible to frost injury than those from northern seed sources. Of particular interest to the western states are the studies on effects of seed source on cold resistance of *Pinus ponderosa* (Squillace and Silen 1962) and *Pseudotsuga menziesii* (Campbell and Sorenson 1973). Another interesting study is that of Rudolph (1964) who showed that the frequency of late-season lammas shoots of *Pinus banksiana* trees grown in Wisconsin increased with more southerly latitude of seed source. The occurrence of late-season lammas shoots increases susceptibility to winter injury (Kozlowski 1971a).

Preparation and Grading of Nursery Stock

In the past, critical age and size of seedlings have been used most commonly in establishing criteria for grading of plantable seedlings. The data on survival and growth obtained from trials with morphological grades often have been exceedingly variable, with results influenced by species, site, year of planting, and other criteria. This was pointed up by Wakeley (1954) in the southern states.

Seedlings originally were graded on presence or absence of secondary needles and winter buds, stem stiffness, proportion of stem having true bark, and relative seedling size. Typical grades were intended to eliminate seedlings too small and too weak-stemmed for probable success.

For a few years morphological grading seemed satisfactory. During the first five years after planting, seedlings of the higher grades showed increased growth and survival over lower grades. However, with wider application, stock graded as plantable often showed low survival. In one outplanting of stock from widely separated and variously fertilized parts of one nursery, great discrepancies in success were noted. For example, in 1934–35 survival varied by 51 percent, and minimum survival was 38 percent. In another planting of 102 lots survival varied by 68 percent with minimum survival 29 percent.

From 1938 to 1941 seedlings were outplanted in the same area. These came from the same nursery as those planted previously, but they had been grown on one small area of uniform soil and under similar intensive management. Survival of such stock was far less variable than in previous trials. Lowest survival among forty-nine lots was 87 percent, and the total range in initial survival of four species of southern pine in three years was only 13 percent.

Subsequent studies cast further doubt on the adequacy of morphological grades as a survival index. Usually grade 1 seedlings showed the best growth, but they generally showed poorer survival than grade 2. Sometimes grade 3 plants, which were considered culls, survived better than grade 1 stock. Many studies since have shown that seedlings of intermediate morphological grades actually have higher capacity for survival than those of highest morphological grades.

Wakeley's experiments demonstrated that critical internal physiological conditions of seedlings, which had not always been easily identified, often greatly outweighed morphological grades in influencing survival and growth after outplanting.

Obviously morphological grades and physiological quality of seedlings may or may not coincide. High physiological quality of southern pine seedlings appears to improve growth and survival through control of internal water balance by insuring adequate water uptake shortly after planting. This appears to be a function of rapid root growth following outplanting. Treatments that reduce physiological quality also interfere with formation of new roots after planting. Survival of transplants often is decreased by treatments that reduce carbohydrate reserves. For example, impeding food accumulation by shading heavily for ten weeks before planting reduced survival of *Pinus palustris* seedlings by 15 to 26 percent, and of *Pinus elliottii* by 56 to 70 percent over controls (Wakeley 1954).

In the northwestern states seedling morphology has been useful as an index of seedling tolerance to environmental stresses. Tall seedlings, large-diameter seedlings, and those with low shoot-root ratios have high field survival potential on adverse sites (Cleary, Greaves, and Owston 1978).

Physiological Quality and Mineral Nutrition. The physiological quality of planting stock is markedly influenced by fertilizer regimes in the nursery. Wakeley (1954) concluded that differences in fertilizer applications were largely responsible for variations in survival among seedling lots from different nurseries, which were similarly graded and outplanted at the same time. Matched plantings of trees from some nurseries survived better than others, regardless of season, site, and even species (table 20). Because planting stock is grown at great density and no crop residues are left after harvest it is tempting to apply fertilizers at very high levels in the nursery. Nevertheless, fertilizer programs must not be aimed only at producing seedlings of maximum size because such seedlings may be of low physiological quality with respect to potential for survival. Unusually heavy fertilizer applications often produce unplantable seedlings with succulent tissues and unbalanced root-shoot ratios. Wilde (1958) among others established standards of nursery soil fertility for producing normally developed planting stock

TABLE 20
VARIATIONS IN SURVIVAL OF GRADED *PINUS ELLIOTTII* SEEDLINGS OBTAINED FROM DIFFERENT NURSERIES AND PLANTED ON SIMILAR SITES

Season and planting location	Seedling Survival (%) from nursery									
	A	B	C	D	E	F	G	H	I	J
1936/37, Alabama					94	39				
1938/39, Mississippi	98	96	80		93	96	57	66		38
1939/40, Mississippi				76	58	54		50	58	
1939/40, Florida				66	24	38		36	27	
1939/40, Louisiana				92	86	77		74	70	

SOURCE: Reprinted, by permission, from Wakeley 1954, modified from table 21.

with high capacity for survival. Such standards obviously will vary with individual species requirements. Wilde also discussed procedures for correcting nursery fertility levels. Usually excessive fertilizer applications, especially nitrogen salts, produce nursery stock of low specific gravity which reflects thin tracheid walls. Such seedlings are very succulent and have low resistance to drought, frost, and perhaps parasitic organisms. Stems of established natural reproduction of *Pinus banksiana* had almost twice the specific gravity of stock of the same age that was grown on heavily fertilized nursery beds (Wilde and Voigt 1948). Pharis and Kramer (1964) found that drought resistance of *Pinus taeda* seedlings was related to the nitrogen regime under which they were grown. Drought resistance was reduced by nitrogen concentrations above optimal levels. The effect of drought on growth was least in nitrogen deficient seedlings, but growth already was at a level so low as to render this decrease of no practical value. Plants grown at nitrogen concentrations that were optimum under normal moisture conditions survived drought best. Lynch et al. (1943) observed that fertilizer treatments that produced healthy appearing nursery plants reduced their survival capacity in plantations. High concentrations of phosphorus reduced survival of both *Pinus taeda* and *Pinus echinata*.

Much further research is needed on physiological characteristics of seedlings that will ensure their success on difficult sites. Detailed experiments will be needed concerning food reserves, growth regulators, nutrient balance, and water stress on response of seedlings after transplanting. In the future simple measurements of stem diameter will be inadequate for a grading system.

A seedling of a given stem diameter range may one year be entirely different from another year, with respect to its physiological vigor which will ultimately govern survival and growth. After a seedling achieves a minimal diameter used in morphological grading systems it can be subjected to nursery treatments that in a short time will influence its replantability. Hence, additional studies are urgently needed that will enable us to identify the internal conditions that are conducive to rapid growth and sufficient water uptake after transplanting to prevent desiccation of tissues. The next obvious need is to work out nursery regimes that will produce such high physiological grades of planting stock. Some criteria of acceptable physiological grades such as color, titration values of root systems, and chemical composition of seedlings have been very useful (Wilde 1958).

Root Pruning and Undercutting. Root pruning after lifting nursery stock and undercutting of roots in nursery beds have often enabled nursery stock to withstand the stresses to which they are exposed after outplanting. In *Pinus radiata* undercutting of roots in nursery beds caused a large increase in translocation of carbohydrates from the foliage to the roots and produced seedlings with only a small amount of soft new shoot growth, a high root-shoot ratio, and a compact mass of fibrous roots. Transplants that received the undercutting treatment survived dry conditions better than control plants partly because the root systems of the former were more efficient in absorption of water (Rook 1969, 1971). Root pruning of containerized plants is a standard nursery practice in many arid-zone nurseries (Goor and Barney 1968).

It appears that root forming substances accumulate near the basal cut surfaces of cuttings. During early stages of root initiation auxins interact with cofactors that are synergists of auxin action. Carlson and Larson (1977) demonstrated that, following pruning of roots of *Quercus rubra* seedlings, both auxins and cytokinins increased rapidly in the roots. The auxin increases that follow such pruning are important because they promote accumulation of carbohydrates and other factors needed for root formation.

Predicting Capacity for Root Growth. Stone and Jenkinson (1971) determined that capacity for rapid root growth following transplanting was important for establishment and growth of *Pinus ponderosa* seedlings. They also concluded that morphological grading of nursery stock had little bearing on whether seedlings survived in the field. Nursery stock of a superior morphological grade often had low capacity to grow new roots when lifted at one time of the year, even when planted in an optimum environment, and a very high capacity when planted a month or two earlier, or later.

Stone and Jenkinson (1971) devised a useful grading system for seedlings that was based on predictability of capacity for root growth. Once nursery cultural practices were standardized and the nursery climate characterized, capacity for root growth could be predicted. The system involved (1) monthly tests of root growth capacity under standardized conditions of seedlings lifted prior to and during the lifting and shipping season, and (2) a cumulative record of the number of hours that air temperatures in the nursery were below 10°C. The system enabled nurserymen to predict seedling root growth for specific planting dates. When used properly, it greatly decreased expensive planting failures caused by the use of seedlings whose expected seasonal root growth was not programmed for the planting site environment.

In the northwestern states seedlings lifted from nursery beds from January to March have the most vigorous root growth after planting. Subsequent cold storage can extend the capacity for maximum root growth beyond March (Clealy, Greaves, and Owston 1978).

"Containerized" Seedlings

During the last decade there has been a surge of interest in producing and planting seedlings in containers. The advantages of doing so include rapid production of planting stock and minimal disturbance of root systems. Since the containers are small, the method makes possible the efficient use of space. Containerized seedlings generally are subjected to intensive cultural techniques resulting in uniform crops of seedlings with specified characteristics. In some regions, containerized seedlings grown in greenhouses with rigid controls of temperature, photoperiod, and CO_2 can be produced in a very short time (ten weeks to a year). Planting of containerized seedlings also is economical. According to Cayford (1972), rates of planting containerized seedlings have varied between 1000 to 2400 plants per man day as against 400 to 900 plants per man day for conventional nursery stock.

Unfortunately, the performance of containerized stock when outplanted has been variable, and sometimes very unsatisfactory. Serious problems with containerized seedlings include slow growth and high mortality of seedlings, lack of frost hardiness, frost heaving, smothering by drifting soil, and damage from pests. Especially needed are regional experiments on size and type of container, type and volume of rooting medium, environmental and cultural conditions for producing stock, methods of site preparation, and methods of planting.

Cleary, Greaves, and Owston (1978) summarized various cultural practices that should be used to produce high-quality containerized seedlings in the northwestern states. They emphasized that irrigation should be heavy and uniform and that fertilizers should be applied frequently and uniformly. Containerized seedlings require a nursery period of six to nine months, with the last half to third of this time devoted to inducing dormancy and frost hardiness. Conditioning for outplanting involves reducing applications of water and nitrogen, reducing daylength, and moving seedlings from greenhouses to shadehouses.

HANDLING OF PLANTING STOCK

All the skill and understanding of producing plantable trees can be undone by improper handling of planting stock. Culling of nursery stock with low potential for survival is desirable. Beyond this, close supervision is needed during transplanting to be sure that nursery plants do not dry out to critical levels. Use of polyethylene packaging is helpful in conserving moisture of planting stock.

Exposure of bare-rooted trees to drying for even short periods of time may have very serious effects on growth and survival, but this varies with species and with the physiological conditions of the plants at the time they are exposed. In one experiment, exposure of nursery stock for as little as four minutes reduced survival. In another experiment, Hermann (1964) exposed 2 + 0 *Pseudotsuga menziesii* seedlings at 90°F and 30 percent relative humidity for periods up to 120 minutes. Survival by November of the year of outplanting was decreased with each added length of exposure. Exposures of thirty minutes influenced small seedlings more than large seedlings but size of plants was unimportant, and survival of both sizes was very low following exposure of thirty minutes or more.

Critical limits of exposure varied considerably with the physiological condition of the nursery stock. Seedlings lifted in the autumn could not survive more than a few minutes of exposure, while those lifted in the winter survived exposure up to thirty minutes. These differences probably were related to root regenerating potential. Prolonged storage also increased susceptibility to exposure. The importance of keeping exposure to a minimum was amply demonstrated. Even if long exposures did not reduce survival they caused outplanted seedlings to grow slowly.

Dipping of roots of planting stock in water immediately on lifting often is beneficial. Mullin (1971) showed that, at times of lifting and planting, dipping of *Picea glauca* roots in water increased survival and improved growth. Increased time of exposure of roots caused significant reduction in growth and survival.

Antitranspirants

Over the years there has been a great deal of interest in maintaining a favorable water balance in trees by applying antitranspirants (antidesiccants) to the leaves. Two main types of antitranspirants are recognized:

1. Film-type antitranspirants which form films on leaves, thereby blocking stomatal pores, or coating the cells inside the leaf with a waterproof film. These include waxes, wax-oil emulsions, higher alcohols, silicones, plastics, latexes, and resins.

2. Metabolic antitranspirants which chemically close stomatal pores. These include succinic acids, phenylmercuric acetate, hydroxysulfonates, the herbicide Atrazine, sodium azide, and phenylhydrazones, and

carbon cyanide. When applied to roots, certain metabolic antitranspirants (e.g., succinic acids) have been reported to increase permeability of roots to water.

We have evaluated the effects of many film-type and metabolic antitranspirants on water loss, photosynthesis, growth, and injury in both angiosperms and gymnosperms. The majority of the antitranspirants tested were toxic to plants and for a very long time. The toxicity, which varied with the antitranspirant, its dosage, and species to which applied, was evident in reduced photosynthesis, altered metabolism, lesions on leaves, chlorosis and browning of leaves, leaf fall, reduced growth, and plant mortality (Kozlowski and Davies 1975). Toxicity sometimes was apparent early and, at other times, late after the compounds were applied. The adverse effects of film-type antitranspirants were particularly long-lived on *Pinus resinosa*. The reduction of plant water loss apparently was the result of the antitranspirant combining with waxes in the stomatal pores and forming impermeable plugs. A number of compounds reduced water loss in pines by as much as 90 percent, but they also drastically reduced photosynthesis and for a very long time. Hence, film-type antitranspirants may be unsuitable for use on gymnosperms that have wax in or around stomatal pores.

If used at all, antitranspirants should be applied conservatively at low dosages to only some of the leaves of a plant so that if these are injured there will be other leaves to carry on the photosynthetic function. Covering only part of the shoot system with an antitranspirant may significantly reduce water loss, maintain a favorable plant water balance, and allow CO_2 exchange to continue at a reasonable level. Such treatment may enable a plant to survive the period between transplanting and resumption of root growth.

Challenges of Research in Forest Biology

FOREST BIOLOGISTS HAVE AN EXCITING CALLING. IT is fascinating to study the marked variability in physiological activity and growth characteristics between temperate and tropical zone trees, gymnosperms and angiosperms, various species and cultivars, and in different parts of the same tree. Under increasing environmental stress, which often is the result of plant competition, the variability in growth of the various components of forest ecosystems reflects differences in competitive ability which play an important role in succession and maintenance of forest types. Unfortunately, we have not yet learned enough about forest ecosystems to manage them to our best advantage. In planning future research related to coping with the impacts of environmental stresses on forest trees I would like to suggest that serious thought be given to the following considerations:

1. Communities of forest trees, as well as individual trees are very complex organisms. In the past we have oversimplified the nature and control of growth. For example, studies on cambial growth were for a long

time concerned largely with xylem production only. Many early researchers assumed that annual xylem production preceded phloem production. However, we now know that in a number of species seasonal phloem differentiation may precede xylem differentiation by as much as several weeks. This emphasizes that overall cambial growth is complicated and should be broken down for study into its several components. More detailed information is needed for different species of trees on maturation of overwintering xylem and phloem cells, division in the cambial sheath to produce xylem and phloem cells, increase in size of cambial derivatives, increase in cell wall thickness, transition from earlywood to latewood, and division of cambial and ray cells to provide for increase in circumference of the cambial cylinder.

Internal control of tree growth has also been oversimplified. For example, effects on cambial growth have often been linked to individual growth regulators, with an important role assigned to auxin. Several early investigators showed that exogenous auxin stimulated cambial division and subsequent differentiation of xylem cells. However, there are problems in interpreting growth responses following application of a single plant growth regulator because it may stimulate metabolic activity to produce other internal growth regulators. It may also influence activity of other growth regulators already present. Although auxin admittedly plays an important role in control of various aspects of cambial activity, normal cambial growth appears to be the end result of balances of several growth regulators (including growth promoters and inhibitors), and interactions among them (Kozlowski 1971b).

2. Considerable basic research and a better climate for it are needed in forest biology. Many forest managers understandably seek rapid solutions to practical problems and have not always been sympathetic to basic research because they considered it an academic exercise not really applicable to field problems. However, history teaches us that basic research leads to widespread application. Medicine made its greatest progress when it began to reinforce practice with basic biochemical research. This also was true in horticulture. Solution of many problems in forestry will depend on a foundation of basic information about physiological processes, life histories, and environmental requirements of various species and genetic materials. Because of the complexity of tree growth we must work long and hard to obtain a pool of basic information needed to solve practical problems. In this connection passage of the McIntyre-Stennis Act in 1963 was one of the best things that ever happened to forestry, to a considerable extent because it placed its stamp of approval on basic as well as applied research. In the last decade, with McIntyre-Stennis support, many significant contributions were made on a number of important problems affecting tree growth. These include breeding for tree form, growth rate, pest resistance, drought resistance, and cold hardiness; simulation modeling; physiology of reproduction; cycling of water and minerals; control of shoot and cam-

bial growth; control of internal water balance in trees; physiological role of mycorrhizae in tree growth; use of systemics in control of insects; chemotherapy of wilt diseases; effects of herbicides, fungicides, and insecticides on physiological processes and growth; formation of plantlets from tissues cultured in vitro; and biological control of insect pests. The progress in all of these areas will have direct or indirect beneficial impacts on our needs for forest products.

3. Some of our most complex biological problems should be investigated with a team-research philosophy. Many forest biologists now agree that more progress would have been made in the past on certain problems if physiologists, soil scientists, geneticists, and morphologists had pooled their talents. In that regard, a 1966 National Academy of Sciences report emphasized that one of the most important needs in forestry research was its close correlation with research in other plant sciences. Team research is particularly important in modeling of forest growth, currently a very fashionable occupation. Often the quality of models has suffered because nonbiologists created them without attention to testing assumptions about biological matters. Interdisciplinary teams, with forest biologists as members, are likely to produce more reasonable and useful growth models than will nonbiologists alone.

4. Forest biologists must be genuinely interested in the effects of heavy cutting on tropical forests. We cannot be too complacent about the prospects of renewability of tropical forest trees on the basis of

what we know about renewability of temperate forests, for the two types of ecosystems are vastly different. It is quite clear that heavy cutting has caused extensive destruction of lowland forests in Borneo, Sumatra, Malaya, and Amazonia. There is much concern that vast areas of tropical forests have been irreparably damaged, before we had adequate knowledge about how they might be managed.

Tropical forests are infinitely more complex and fragile ecosystems than are temperate-zone forests. Most tropical forests are closed ecosystems that exist on a very small nutrient budget. In contrast to the situation in temperate forests, most of the cycling of nutrient capital of tropical forests is in the plants themselves, particularly in leaves and twigs. The nutrients in dead wood and leaves and in excretions and dead bodies of animals are rapidly released by the action of decomposed organisms. Once dead material is broken down, the minerals are rapidly absorbed by roots of trees and other plants. Hence the soils are relatively infertile, but recycling is rapid and efficient and there is little nutrient loss from the system. The high efficiency of the mineral cycle is shown by low concentrations of mineral nutrients in rivers that drain tropical forest areas.

Succession in tropical forests is very slow. Clearings in primary tropical forests soon are covered with a dense growth of weeds, shrubs, vines, and young trees. Soon rapid-growing, short-lived trees with a life span of less than twenty years take over. These are succeeded by intermediate, slower-growing and longer-lived species. Succession proceeds slowly until finally the pioneer species are reestablished. The time of clearing to reestablishment of primary forest may take hundreds of years. Often such succession is further interrupted by harvesting of secondary forest species for firewood and by clearing for cultivation. This added interruption results in additional loss of nutrients from the soil so the land cannot even support a secondary forest. It then supports highly flammable savanna, grasses, bamboo, and ferns. The result is indefinite postponement of establishment of primary forest.

Tropical forests, what remains of them, are superb laboratories for studies of the complex principles of ecological balances and renewability of forest trees. Regrettably there has been overcutting in tropical forests in recent years without regard to the impact of such cutting on regeneration. We really need to know much more about the amount of imposed stress that various tropical forest ecosystems can tolerate without losing their capacity to recover. More quantitative data are needed on rates of productivity, decomposition, and mineral cycling in order to evaluate potential regeneration of tropical forests. Additional information is needed on succession to determine how management programs might affect recovery of disturbed tropical forests. We should also try to understand processes of undisturbed forests in order to assess the effect of disturbance against a standard. We also need more data on microclimate, plants, animals, and soils as they change following different levels of disturbance in tropical ecosystems.

5. Controlled environments are very useful in solving field problems. Many problems exist in evaluating the impact of environment on tree growth in the field (Kozlowski 1974). For example, light, temperature, humidity, and other factors are so interdependent that a change in one alters the others so it becomes extremely difficult to evaluate the effect of individual factors, or interactions among them, on tree growth. Furthermore, it is virtually impossible in the field to reproduce precisely the environment of any one day on succeeding days. Hence, we should give careful thought to the values of phytotrons and biotrons. The many carefully controlled environmental combinations that are available in such facilities make it possible to study environmental factor action and interactions with high accuracy and within a short time. Phytotrons offer several advantages in research over field conditions, greenhouses, or even small groups of controlled environment chambers. A phytotron can be used to dissect or construct a given environment. Growth responses over a wide range of given environmental factors can be obtained concurrently. Interaction of two or more factors can be studied while holding other factors relatively constant. The great value of biotrons and phytotrons is that they decrease variability and increase reproducibility. Biotrons are superb facilities for research on (1) effects of environmental factors (e.g., light intensity, photoperiod, temperature, soil water, humidity, wind, etc.) and

their interactions on physiology of growth and development; (2) study of adaptation and acclimation; (3) screening plant material for special phenological characteristics; (4) interactions between plants and other organisms; (5) effects of applied and naturally occurring biocides on trees under varying environmental conditions; and (6) determination of environmental factors that are likely to be significant in the field and therefore worth measuring. Information so obtained can materially reduce the extent and cost of field research.

Biotrons should be used to answer important questions and not merely for growing plants. Although the costs of biotron research are high, they often are overestimated when compared with costs of field research. Biotron research may cost less than field research because the smaller number of plants necessary per sample in the biotron than in the field reduces the cost of laboratory analysis. Furthermore, in comparing costs, accounting generally is not made of the high price of maintaining and managing experimental areas, costs of field travel, costs of damage to field experiments by weather, insects, and disease (requiring repeating field experiments), cost of delays in waiting for specific weather conditions, etc. (Kramer, Slatyer, and Hellmers 1972).

A few examples of the kinds of problems that can be investigated rapidly and efficiently in biotrons include the following:

Biotrons can be used to select climatic regimes in which to field-test new species and varieties or to assist in selection of plants for use in specific climates. Biotrons are especially useful to predetermine environmental regimes for producing containerized seedlings for particular sites. Hellmers (1967) grew *Picea engelmannii* seedlings under thirty combinations of day and night temperature. He found that night temperature was the most important facet of the temperature regime. A night temperature of 23° C, with a 15° to 35 °C range of day temperatures, produced seedlings in seven months that were comparable to five-year-old nursery-grown stock. Owston and Kozlowski (1978) grew containerized *Pseudotsuga menziesii, Abies nobilis,* and *Picea sitchensis* seedlings (seed collected from different elevations in the Pacific Northwest) in the University of Wisconsin Biotron under conditions that simulated hot, warm, and cool growing seasons in greenhouses near Corvallis, Oregon. A simulated growing season from late April through early October was divided into the germination period (in standard growth chambers) and six growth periods in the Biotron. Maximum temperatures for the warm regime were based on average monthly maxima for Corvallis, Oregon. Maxima for the hot and cool regimes were set 6° or 8° C higher or lower than the corresponding maximum for the warm regime.

Seeds were germinated in standard growth rooms in mixtures of peat moss and vermiculite in 3-cm-wide × 13-cm-deep cavities of a polystyrene block. Seedlings were then grown in each of the three Biotron regimes. Seedlings were fertilized and watered regularly. Growth data were taken periodically for five months on sample seedlings from each regime. Finally sample plants were subjected to freezing tests.

Growth of seedlings varied among species and seed sources. Dry weight increments were highest for plants of the warm regime, except that they were highest in the hot regime for one *Pseudotsuga menziesii* source from the southern Oregon Cascades, an area with hot summers.

The freezing test was conducted when the seedlings were not completely hardened to frost. It was performed, however, at a time when fall planting begins on some sites in the Pacific Northwest. Of particular interest was the fact that seedlings grown in the hot regime were injured much more by freezing than seedlings from the other two regimes. Differences in susceptibility to frost between seedlings of the warm and cool regimes were less pronounced and different among species.

These experiments emphasized that much more research is needed under controlled conditions to determine the best greenhouse regimes for growing containerized seedlings of different species and provenances for planting on individual sites. Generally, careful attention should be given to inducing frost hardiness in containerized plants, at least in the latter part of the growing season. This is particularly true for plants scheduled for fall planting.

6. Forest biologists should become more involved in public relations and in decision-making pertaining to biological research and administration. The general public is understandably becoming very concerned

about possible damage to the environment. Recognizing the public's right to know what is happening to our natural resources, forest biologists can perform a service by *accurately* interpreting the impact of environmental stresses on forest ecosystems. They can also stimulate support for eliminating man-made stresses, such as pollution, at the source. Interpreting what pollutants do to ecosystems will be useful in that regard. Regrettably, many instant ecologists have emerged in the last decade and have pushed panic buttons without adequate reinforcing data. I prefer the sensible public relations work of well-trained forest biologists and, in the long run, I believe the public will also.

Finally, forest biologists should assert themselves so as to play a forceful role in planning of environmental research. In the past many excellent biologists have tended to abdicate from that responsibility, and, by default, decisions that affect biological research have much too often been made, and not always well, by nonbiologists.

In the immediate future the solution of environmental problems in forest ecosystems will be of paramount importance if we are to sustain life and its quality. Forest biologists will have a vital role to play in the solution of those problems.

Appendix 1

HERBICIDES MENTIONED

Atrazine	2-chloro-4-ethylamino-6-isopropylamino-s-triazine
CDAA	2-chloro-N,N-diallylacetamide
CDEC	2-chloroallyl diethyldithiocarbamate
2,4-D	2,4-dichlorophenoxyacetic acid
2,4,5-T	2,4,5-trichlorophenoxyacetic acid
DCPA	dimethyl-2,3,5,6-tetrachloroterephthalate
EPTC	ethyl N,N-di-n-propylthiolcarbamate
Fenuron	1,1-dimethyl-3-phenylurea
Ipazine	2-chloro-4-diethylamino-6-isopropylamino-s-triazine
Monuron	3-(p-chlorophenyl)-1,1-dimethylurea
NPA	N-1-naphthylphthalamic acid
Picloram	4-amino-3,5,6-trichloropicolinic acid
Prometone	2-methoxy-4,6-bis(isopropylamino)-s-triazine
Prometryne	2,4-bis(isopropylamino)-6-methylmercapto-s-triazine
Propazine	2-chloro-4,6-bis(isopropylamino)-s-triazine
Sesone	sodium 2,4-dichlorophenoxyethyl sulfate
Simazine	2-chloro-4,6-bis(ethylamino)-s-triazine

Appendix 2

Scientific and Common Names of Species Mentioned

SCIENTIFIC NAME	COMMON NAME
Abies amabilis (Dougl.) Forbes	Pacific silver fir
Abies balsamea (L.) Mill.	Balsam fir
Abies concolor (Gord. and Glend.) Lindl.	White fir
Abies grandis (Dougl.) Lindl.	Grand fir
Abies lasiocarpa (Hook.) Nutt.	Sub-alpine fir
Abies nobilis (Dougl.) Lind. (see *Abies procera* Rehd.)	
Abies procera Rehd.	Noble fir
Acer circinatum Pursh.	Vine maple
Acer grandidentatum Nutt.	Bigtooth maple
Acer negundo L.	Box elder
Acer platanoides L.	Norway maple
Acer rubrum L.	Red maple
Acer saccharinum L.	Silver maple
Acer saccharum Marsh.	Sugar maple
Ailanthus altissima (Mill.) Swingle	Ailanthus
Arbutus menziesii Pursh.	Pacific madrone
Betula alleghaniensis Britton	Yellow birch

Betula papyrifera Marsh. — Paper birch
Betula pendula Roth — European white birch
Betula populifolia Marsh. — Gray birch
Calluna vulgaris (L.) Hull — Heather
Carya cordiformis (Wangenh.) K. Koch — Bitternut hickory
Celtis occidentalis L. — Hackberry
Cercis canadensis L. — Eastern redbud
Citrus spp. — Citrus
Cornus florida L. — Flowering dogwood
Cornus Nuttallii Audub. — Mountain dogwood
Cornus stolonifera Michx. — Red-osier dogwood
Eucalyptus camaldulensis Dehn. (*E. rostrata* Schlect.) — River red gum
Eucalyptus incrassata Labill. — Lerp mallec
Eucalyptus rostrata Schlect. (see *E. camaldulensis* Dehn.)
Eucalyptus socialis F. Muell. — Eucalyptus
Fagus sylvatica L. — European beech
Fraxinus americana L. — White ash
Fraxinus pennsylvanica Marsh. — Green ash
Gingko biloba L. — Gingko
Gleditsia triacanthos L. — Honey locust
Gymnocladus dioicus (L.) K. Koch — Kentucky coffee tree
Ilex opaca Ait. — American holly
Juglans nigra L. — Black walnut
Juglans regia L. — English walnut
Juniperus occidentalis Hook. — Western juniper
Kalmia angustifolia L. — Sheep laurel
Larix decidua Mill. — European larch
Larix laricina (DuRoi) K. Koch — Tamarack
Larix leptolepis Murr. — Japanese larch
Larix occidentalis Nutt. — Western larch
Libocedrus decurrens Torr. — Incense-cedar
Ligustrum obtusifolium Siebold & Zucc. — Regel's privet

Liquidambar styraciflua L.	Sweetgum
Liriodendron tulipifera L.	Yellow poplar
Magnolia grandiflora L.	Southern magnolia
Malus x *robusta* (Carriere) Rehd.	Apple
Nyssa aquatica L.	Water tupelo
Nyssa silvatica Marsh.	Black tupelo
Physocarpus capitatus (Pursh) O. Kuntze	Ninebark
Picea abies (L.) Karst.	Norway spruce
Picea engelmannii Parry	Engelmann spruce
Picea glauca (Moench) Voss	White spruce
Picea mariana (Mill.) B.S.P.	Black spruce
Picea pungens Engelm.	Blue spruce
Picea sitchensis (Bong.) Carr.	Sitka spruce
Pinus attenuata Lemm.	Knobcone pine
Pinus banksiana Lamb.	Jack pine
Pinus caribaea Moulet	Cuban pine
Pinus contorta Dougl.	Lodgepole pine
Pinus coulteri D. Don	Coulter pine
Pinus echinata Mill.	Shortleaf pine
Pinus edulis Engelm.	Pinyon pine
Pinus elliottii Engelm.	Slash pine
Pinus flexilis James	Limber pine
Pinus jeffreyi Grev. & Balf.	Jeffrey pine
Pinus lambertiana Dougl.	Sugar pine
Pinus monticola Dougl.	Western white pine
Pinus nigra Arnold	Austrian pine
Pinus ponderosa Laws	Ponderosa pine
Pinus radiata D. Don	Monterey pine
Pinus resinosa Ait.	Red pine
Pinus rigida Mill.	Pitch pine
Pinus sabiniana Dougl.	Digger pine
Pinus silvestris L.	Scotch pine

Pinus strobus L. — Eastern white pine
Pinus taeda L. — Loblolly pine
Pinus virginiana Mill. — Virginia pine
Platanus occidentalis L. — American sycamore
Populus alba L. — White poplar
Populus balsamifera L. — Balsam poplar
Populus deltoides Bartr. — Eastern cottonwood
Populus grandidentata Michx. — Bigtooth aspen
Populus tremuloides Michx. — Trembling aspen
Populus trichocarpa Torr. & Gray — Black cottonwood
Prunus pumila L. — Sand cherry
Prunus serotina Ehrh. — Black cherry
Prunus virginiana L. — Chokecherry
Pseudotsuga menziesii (Mirb.) Franco — Douglas-fir
Quercus alba L. — White oak
Quercus borealis maxima (Marsh.) Ashe, (see *Q. rubra* L.)
Quercus coccinea Muenchh. — Scarlet oak
Quercus ellipsoidalis E.J. Hill — Northern pin oak
Quercus gambelii Nutt. — Gambel oak
Quercus imbricaria Michx. — Shingle oak
Quercus macrocarpa Michx. — Bur oak
Quercus palustris Muenchh. — Pin oak
Quercus robur L. — English oak
Quercus rubra L. *(Quercus borealis, maxima* [Marsh.] Ashe) — Eastern red oak
Quercus velutina Lam. — Black oak
Quercus virginiana Mill. — Live oak
Robinia pseudoacacia L. — Black locust
Rubus parviflorus Nutt. — Thimbleberry
Salix atrocinerea Brot. — Gray Willow
Salix interior Rowlee — Sandbar willow

Appendix 2

Salix nigra Marsh. — Black willow
Salix pellita Anderss. — Willow
Sambucus racemosa L. — European red elder
Sequoia sempervirens (D. Don) Endl. — Redwood
Sequoiadendron giganteum (Lindl.), Buchh. — Giant sequoia
Symphoriocarpus albus (L.) Blake — Snowberry
Syringa vulgaris L. — Lilac
Thuja occidentalis L. — Northern white-cedar
Thuja plicata Don — Western redcedar
Tilia americana L. — American basswood
Tilia cordata Mill. — Little leaf linden
Tsuga canadensis (L.) Carr. — Eastern hemlock
Tsuga heterophylla Vent. — White basswood
Tsuga mertensiana (Bong.) Carr. — Mountain hemlock
Ulmus americana L. — American elm
Ulmus parvifolia Jacq. — Chinese elm

Literature Cited

Addicott, F. T. 1970. Plant hormones in the control of abscission. *Biol. Rev. Cambridge Phil. Soc.* 45:485–524.

———, and R. S. Lynch. 1955. Physiology of abscission. *Ann. Rev. Plant Physiol.* 6:211–38.

Alben, A. O. 1958. Waterlogging of subsoil associated with scorching and defoliation of Stuart pecan trees. *Proc. Amer. Soc. Hort. Sci.* 72:219–23.

Alberte, R. S., P. R. McClure, and J. P. Thornber. 1976. Photosynthesis in trees: Organization of chlorophyll and photosynthetic unit size in isolated gymnosperm chloroplasts. *Plant Physiol.* 58:341–44.

Albertson, F. W., and J. E. Weaver. 1945. Injury and death or recovery of trees in prairie climate. *Ecol. Monogr.* 15:393–433.

Alden, J., and R. K. Hermann. 1971. Aspects of the cold-hardiness mechanism in plants. *Bot. Rev.* 37:37–142.

Azevedo, J., and D. L. Morgan. 1974. Fog precipitation in coastal California forests. *Ecology* 55:1135–41.

Barber, H. N., and W. D. Jackson. 1957. Natural selection in action in *Eucalyptus. Nature* 179:1267–79.

Barrs, H. D. 1968. Determination of water deficits in plant tissues. In *Water Deficits and Plant Growth,* vol. 1, ed.

T. T. Kozlowski, pp. 235–368. New York: Academic Press.

Bartholomew, E. T. 1926. Internal decline of lemons, 3: Water deficit in lemon fruits caused by excessive leaf evaporation. *Amer. J. Bot.* 13:102–17.

Bormann, F. H. 1966. The structure, function, and ecological significance of root grafts in *Pinus strobus* L. *Ecol. Monogr.* 36:1–26.

———, and T. T. Kozlowski. 1962. Measurements of tree ring growth with dial gage dendrometers and vernier tree ring bands. *Ecology* 43:289–94.

Boyer, J. S. 1973. Response of metabolism to low water potentials in plants. *Phytopathology* 63:466–72.

Boyle, J. R. 1975. Nutrients in relation to intensive culture of forest crops. *Iowa State J. Res.* 49:297–303.

Braekke, F. H. 1976. Impact of acid precipitation on forest and freshwater ecosystems in Norway. *Res. Rept. FR6/76.* Ås, Norway: SNSF Proj.

———, and T. T. Kozlowski. 1975. Shrinkage and swelling of stems of *Pinus resinosa* and *Betula papyrifera* in northern Wisconsin. *Plant and Soil* 43:387–410.

Braun, G. 1977a. Causes and criteria of resistance to air pollution in Norway spruce, 1. Morphological and anatomical resistance. *Eur. J. Forest Path.* 7:23–43.

———. 1977b. Causes and criteria of resistance to air pollution in Norway spruce. II. Reflexive resistance. *Eur. J. Forest Path.* 7:129–52.

———. 1977c. Causes and criteria of resistance to air pollution in Norway spruce, 3. Tolerance of toxic materials ("internal" resistance). *Eur. J. Forest Path.* 7:236–49.

———. 1977d. Causes of resistance against air pollution in Norway spruce and conclusions in respect to resistance breeding. *Forstwiss. Centralbl.* 96:62–66.

Bryson, R. A., and T. J. Murray. 1977. *Climates of Hunger.* Madison, Wisconsin: Univ. of Wisconsin Press.

Buell, M. F., J. A. Small, and C. D. Monk. 1961. Drought

effect on radial growth of trees in the William L. Hutcheson Memorial Forest (New Jersey). *Bull. Torrey Bot. Club* 88:176–80.

Burrows, W. J., and D. J. Carr. 1969. The effects of flooding of the root system of sunflower plants on the cytokinin content in the xylem sap. *Physiol. Plant.* 22:1105–12.

Campbell, R. K., and F. C. Sorenson. 1973. Cold acclimation in seedling Douglas-fir related to phenology and provenance. *Ecology* 54:1148–51.

Carlson, W. C., and M. M. Larson. 1977. Changes in auxin and cytokinin activity in roots of red oak, *Quercus rubra*, seedlings during lateral root formation. *Physiol. Plant.* 41:162–66.

Cayford, G. H. 1972. Container planting systems in Canada. *Forestry Chron.* 48:235–39.

Chaney, W. R., and T. T. Kozlowski. 1969a. Diurnal expansion and contraction of leaves and fruits of English Morello cherry. *Ann. Bot.* 33:691–99.

———. 1969b. Seasonal and diurnal expansion and contraction of *Pinus banksiana* and *Picea glauca* cones. *New Phytol.* 68:873–82.

———. 1971. Water transport in relation to expansion and contraction of leaves and fruits of Calamondin orange. *J. Hort. Sci.* 46:71–81.

Clausen, J. J., and T. T. Kozlowski. 1967. Seasonal growth characteristics of long and short shoots of tamarack. *Can. J. Bot.* 45:1643–51.

Cleary, B. D., R. D. Greaves, and P. W. Owston. 1978. Seedlings. In *Regenerating Oregon's Forests,* eds. B. D. Cleary, R. D. Greaves, and R. K. Hermann, pp. 63–97. Corvallis, Oregon: Oregon State Univ., School of Forestry.

Clements, J. R. 1970. Shoot responses of young red pine to watering applied over two seasons. *Can. J. Bot.* 48:75–80.

Constantinidou, H. A., T. T. Kozlowski, and K. Jensen.

1976. Effect of sulfur dioxide on *Pinus resinosa* seedlings in the cotyledon stage. *J. Environ. Qual.* 5:141–44.

Costonis, A. C. 1970. Acute foliar injury of eastern white pine induced by sulfur dioxide and ozone. *Phytopathology* 60:994–99.

Daubenmire, R. 1978. *Plant Geography.* New York: Academic Press.

Davies, W. J., and T. T. Kozlowski. 1974. Stomatal responses of five woody angiosperms to light intensity and humidity. *Can. J. Bot.* 52:1525–34.

———, T. T. Kozlowski, W. R. Chaney, and K. Lee. 1973. Effects of transplanting on physiological responses and growth of shade trees. *Int. Shade Tree Conf. Proc.* 48 (1972):22–30.

———, T. T. Kozlowski, and J. Pereira. 1974. Effect of wind on transpiration and stomatal aperture of woody plants. In *Mechanisms of Regulation of Plant Growth,* eds. R. L. Bieleski et al., pp. 433–38. Roy. Soc. New Zealand, bull. 12.

Davis, D. D., and H. D. Gerhold. 1976. Selection of trees for tolerance of air pollutants. In *Better Trees for Metropolitan Landscapes,* eds. F. S. Santamour, H. D. Gerhold, and S. Little, pp. 61–66. USDA Forest Service General Tech. Rept. NE-22.

De Candolle, M. A.-P. 1832. *Physiologie Vegetale,* vol. 3. Bechet Jeune. Paris: Lib. Fac. Med.

del Moral, R., and R. G. Cates. 1971. Allelopathic potential of the dominant vegetation of western Washington. *Ecology* 52:1030–37.

———, and C. H. Muller. 1970. The allelopathic effects of *Eucalyptus camaldulensis. Amer. Midl. Natur.* 83:254–82.

Dickmann, D. I., and T. T. Kozlowski. 1969. Seasonal growth patterns of ovulate strobili of *Pinus resinosa* in central Wisconsin. *Can. J. Bot.* 47:839–48.

Dimock, E. W. 1964. Simultaneous variations in seasonal

height and radial growth of young Douglas-fir. *J. Forestry* 62:252–55.

Dobbs, R. C., and D. R. M. Scott. 1971. Distribution of diurnal fluctuations in stem circumference of Douglas-fir. *Can. J. For. Res.* 1:80–83.

Dochinger, L. S., and T. A. Seliga. 1975. Acid precipitation and the forest ecosystem. *APCA J.* 25:1103–5.

Durzan, D. J. 1971. Free amino acids as affected by light intensity and the relation of responses to the shade tolerance of white spruce and shade intolerance of jack pine. *Can. J. For. Res.* 1:131–40.

Duvdevani, A. 1964. Dew in Israel and its effect on plants. *Soil Sci.* 98:14–21.

Eis, S. 1972. Root grafts and their silvicultural implications. *Can. J. For. Res.* 2:111–20.

Eiten, G. 1972. The cerrado vegetation of Brazil. *Bot. Rev.* 38:201–341.

Evans, G. C. 1956. An area survey method of investigating the distribution of light intensity on woodlands with particular reference to sunflecks. *J. Ecol.* 44:391–428.

Fisher, R. F. 1976. Allelopathic interference among plants, 1. Ecological significance. *Proceedings Fourth North Amer. Forest Biology Workshop.* Syracuse, New York: State Univ. Coll. Environ. Sci. and Forestry, pp. 73–92.

Franklin, J. F., W. H. Moir, G. W. Douglas, and C. Wiberg. 1971. Invasion of subalpine meadows by trees in the Cascade Range, Washington and Oregon. *Arctic and Alpine Res.* 3:215–24.

Fritschen, L. J., and P. Doraiswamy. 1973. Dew: An addition to the hydrologic balance of Douglas-fir. *Water Resources Res.* 9:891–94.

Fritts, H. C., D. G. Smith, J. W. Cardis, and C. A. Budelsky. 1965. Tree-ring characteristics along a vegetation gradient in northern Arizona. *Ecology* 46:393–401.

Furr, J. R., and C. A. Taylor. 1939. Growth of lemon fruits

in relation to moisture content of the soil. *USDA Tech. Bull.* 640.

Gill, C. J. 1970. The flooding tolerance of woody species: A review. *Forestry Abstr.* 31:671–88.

Gindel, I. 1965. Irrigation of plants with atmospheric water within the desert. *Nature* 207:1173–75.

Glerum, C. 1976. Forest hardiness of forest trees. In *Tree Physiology and Yield Improvement,* eds. M. G. R. Cannell and F. T. Last, pp. 403–20. New York: Academic Press.

Glock, W. S., R. A. Studhalter, and S. R. Agerter. 1960. Classification and multiplicity of growth layers in the branches of trees at the extreme lower forest border. *Smithsonian Misc. Publ.* 4421.

Goor, A. Y., and C. W. Barney. 1968. *Forest Tree Planting in Arid Zones.* New York: Ronald Press.

Haberland, F. P., and S. A. Wilde. 1961. Influence of thinning of red pine plantation on soil. *Ecology* 42:584–86.

Hall, D. M., A. I. Matus, J. A. Lamberton, and H. N. Barber. 1965. Intraspecific variation in wax on leaf surfaces. *Austr. J. Biol. Sci.* 18:323–32.

Hall, T. F., and G. E. Smith. 1955. Effects of flooding on woody plants, West Sandy dewatering project, Kentucky Reservoir. *J. Forestry* 53:281–85.

Hallaway, H. M., and D. J. Osborne. 1969. Ethylene: A factor in defoliation by auxins. *Science* 163:1067–68.

Heide, O. M. 1974. Growth and dormancy in Norway spruce ecotypes *(Picea abies),* 1: Interaction of photoperiod and temperature. *Physiol. Plant.* 30:1–12.

Hellmers, H. 1967. Controlled environments for measuring early responses of tree seedlings to temperature. *14th IUFRO Congress Proc.,* 491–514.

Hermann, R. K. 1964. Effects of prolonged exposure of

roots on survival of 2–0 Douglas-fir seedlings. *J. Forestry* 62:401–3.

Hinckley, T. M., and D. M. Bruckerhoff. 1975. The effects of drought on water relations and stem shrinkage of *Quercus alba. Can. J. Bot.* 53:62–72.

Hook, D. D., C. L. Brown, and P. P. Kormanik. 1970. Lenticels and water root development of swamp tupelo under various flooding conditions. *Bot. Gaz.* 13:217–21.

———. 1971. Inductive flood tolerance in swamp tupelo (*Nyssa sylvatica* var. *biflora* [Walt.] Sarg.). *J. Exptl. Bot.* 22:178–89.

Hook, H. D., C. L. Brown, and R. H. Wetmore. 1972. Aeration in trees. *Bot. Gaz.* 133:443–54.

Horsley, S. B. 1977. Allelopathic inhibition of black cherry by fern, grass, goldenrod, and aster. *Can. J. For. Res.* 7:205–16.

Houston, D. B., and L. S. Dochinger. 1977. Effects of ambient air pollution on cone, seed and pollen characteristics in eastern white and red pines. *Environ. Pollut.* 12:1–5.

Houston, O. 1970. Physiological and genetic responses of *Pinus strobus* L. clones to sulfur dioxide and ozone exposures. Ph.D. dissertation. Madison, Wisconsin: Univ. Wisconsin.

Howe, J. P. 1968. Influence of irrigation on ponderosa pine. *Forest Prod. J.* 18:84–93.

Hsiao, T. C. 1973. Plant responses to water stress. *Ann. Rev. Plant Physiol.* 24:519–70.

Jarvis, P. G. 1975. Water transfer in plants. In *Heat and Mass Transfer in the Environment of Vegetation,* ed. D. A. De Vries, pp. 369–94. Seminar of Int. Center for Heat and Mass Transfer, Dubrovnik. Washington, D. C.: Scripta Book Co.

———. 1976. The interpretations of the variations in leaf water potential and stomatal conductance found in canopies in the field. *Phil. Trans. Roy. Soc. London B.* 273:593–610.

Jeffree, C. E., R. P. C. Johnson, and P. G. Jarvis. 1971. Epicuticular wax in the stomatal antechamber of Sitka spruce and its effects on the diffusion of water vapor and carbon dioxide. *Planta* 98:1–10.

Jensen, K. F., and T. T. Kozlowski. 1970. Absorption and translocation of sulfur dioxide by seedlings of four forest tree species. *J. Environ. Qual.* 4:379–81.

Jonsson, B., and R. Sundberg. 1972. Has the acidification by atmospheric pollution caused a growth reduction in Swedish forests? Note no. 20, *Dept. of Forest Yield Res.* Stockholm, Sweden: Royal Coll. Forestry.

Kawase, M. 1972. Effect of flooding on ethylene concentration in horticultural plants. *J. Amer. Soc. Hort. Sci.* 97: 584–88.

———. 1974. Role of ethylene in induction of flooding damage in sunflower. *Physiol. Plant.* 31:29–38.

Kimmins, J. P. 1977. Evaluation of the consequences for future tree productivity of the loss of nutrients in whole-tree harvesting. *Forest Ecol. and Mgmt.* 1:169–83.

Kincer, J. B. 1934. Data on the drought. *Science* 80:179.

Kozlowski, T. T. 1968a. *Water Deficits and Plant Growth*, vol. 1: *Development, Control, and Measurement.* New York: Academic Press.

———. 1968b. *Water Deficits and Plant Growth*, vol. 2: *Plant Water Consumption and Response.* New York: Academic Press.

———. 1969. Tree physiology and forest pests. *J. Forestry* 69:118–22.

———. 1971a. *Growth and Development of Trees*, vol. 1: *Seed Germination, Ontogeny, and Shoot Growth.* New York: Academic Press.

———. 1971b. *Growth and Development of Trees*, vol. 2:

Cambial Growth, Root Growth, and Reproductive Growth. New York: Academic Press.

———. 1972a. Physiology of water stress. In *Wildland Shrubs: Their Biology and Utilization. USDA Forest Service. Gen. Tech. Rept. INT-1:* 229–44. Ogden, Utah.

———. 1972b. Shrinking and swelling of plant tissues. In *Water Deficits and Plant Growth,* vol. 3, ed. T. T. Kozlowski, pp. 1–64. New York: Academic Press.

———. 1972c. *Water Deficits and Plant Growth,* vol. 3: *Plant Responses and Control of Water Balance.* New York: Academic Press.

———. 1974. The role of educational institutions in forest biology research. In *Proceedings of the Third North American Forest Biology Workshop,* eds. C. P. P. Reid and G. H. Fechner, pp. 275–317. Ft. Collins, Colorado: Colorado State University.

———. 1976a. Susceptibility of young tree seedlings to environmental stresses. *Amer. Nurseryman* 144, no. 11: 12, 13, 55–59.

———. 1976b. *Water Deficits and Plant Growth,* vol. 4: *Soil Water Measurement, Plant Responses, and Breeding for Drought Resistance.* New York: Academic Press.

———. 1978. *Water Deficits and Plant Growth,* vol. 5: *Water and Plant Diseases.* New York: Academic Press.

———, and Borger, G. A. 1971. Effect of temperature and light intensity early in ontogeny on growth of *Pinus resinosa* seedlings. *Can. J. For. Res.* 1: 57–65.

———, and J. J. Clausen. 1965. Changes in moisture contents and dry weights of buds and leaves of forest trees. *Bot. Gaz.* 126:20–26.

———, and J. J. Clausen. 1966. Shoot growth characteristics of heterophyllous woody plants. *Can. J. Bot.* 44: 827–43.

———, and J. C. Cooley. 1961. Root grafting in northern Wisconsin. *J. Forestry* 59:105–7.

———, and W. J. Davies. 1975. Control of water balance in transplanted trees. *J. Arboriculture* 1:1–10.

———, W. J. Davies, and S. D. Carlson. 1974. Transpiration rates of *Fraxinus americana* and *Acer saccharum* leaves. *Can. J. For. Res.* 4:259–67.

———, and T. E. Greathouse. 1970. Shoot growth characteristics of tropical pines. *Unasylva* 24:1–10.

———, and T. Keller. 1966. Food relations of woody plants. *Bot. Rev.* 32:293–382.

———, and J. E. Kuntz. 1963. Effect of Simazine, Atrazine, Propazine, and Eptam on growth of pine seedlings. *Soil Sci.* 95:164–74.

———, and S. Sasaki. 1970. Effects of herbicides on seed germination and development of young pine seedlings. In *Proc. Int. Symp. on Seed Physiology of Woody Plants*. Poznan, Poland (1968), pp. 19–24.

———, and J. H. Torrie. 1965. Effect of soil incorporation of herbicides on seed germination and growth of pine seedlings. *Soil Sci.* 100: 139–46.

———, J. H. Torrie, and P. E. Marshall. 1973. Predictability of shoot length from bud size in *Pinus resinosa* Ait. *Can. J. For. Res.* 3:34–38.

———, and C. H. Winget. 1964. Diurnal and seasonal variation in radii of tree stems. *Ecology* 45:149–55.

———, C. H. Winget, and J. H. Torrie. 1962. Daily radial growth of oak in relation to maximum and minimum temperature. *Bot. Gaz.* 124:9–17.

Kramer, P. J. 1950. Soil aeration and tree growth. *Proc. 26th Int. Shade Tree Conf.*, pp. 51–58.

———. 1956. The role of physiology in forestry. *Forestry Chron.* 32:297–308.

———, and T. T. Kozlowski. 1960. *Physiology of Trees*. New York: McGraw-Hill.

———. R. O. Slatyer, and H. Hellmers. 1972. Phytotrons and environmental physiology. *Nature and Resources* 8(4):13–16. Paris: UNESCO.

Kraus, J. F. , and S. H. Spurr. 1961. Relationship of soil

moisture to the springwood-summerwood transition in southern Michigan red pine. *J. Forestry* 59:510–11.
Larcher, W. 1975. *Physiological Plant Ecology.* New York: Springer-Verlag.
Lassoie, J. P., and D. R. M. Scott. 1977. Water relations of vine maple in a Douglas-fir stand. In *Proceedings of the Symposium on Terrestrial and Aquatic Ecological Studies of the Northwest,* eds. R. D. Andrews, III, R. L. Carr, F. Gibson, B. Z. Lang, R. A. Soltero, and K. C. Swedberg, pp. 23–37. Cheney, Washington: EWSC Press.
Leopold, A. C., and P. E. Kriedemann. 1975. *Plant Growth and Development.* New York: McGraw-Hill.
Lerner, R. H., and M. Evenari. 1961. The nature of the germination inhibitor present in leaves of *Eucalyptus rostrata. Physiol. Plant.* 14:221–29.
Levitt, J. 1972. Responses of Plants to Environmental Stresses. New York: Academic Press.
Leyton, L., and I. P. Armitage. 1968. Cuticle structure and water relations of the needles of *Pinus radiata* (D. Don.). *New Phytol.* 67:31–38.
———, and B. E. Juniper. 1963. Cuticle structure and water relations of pine needles. *Nature* 198: 770–71.
Likens, G. E., and F. H. Bormann. 1974. Acid rain: A serious regional environmental problem. *Science* 184: 1176–79.
Lister, G. R., V. Slankis, G. Krotkov, and C. D. Nelson. 1967. Physiology of *Pinus strobus* L. seedlings grown under high or low soil moisture conditions. *Ann. Bot.* 31:121–32.
Logan, K. T. 1971. Monthly variations in photosynthetic rate of jack pine provenances in relation to their height. *Can. J. For. Res.* 1:256–61.
Lopushinsky, W. 1969. Stomatal closure in conifer seedlings in response to leaf moisture stress. *Bot. Gaz.* 130: 258–63.
Lynch, D. W., W. C. Davis, L. R. Roof, and C. F. Korstian.

1943. Influence of nursery fungicide-fertilizer treatments on survival and growth in a southern pine plantation. *J. Forestry* 41:411–13.

Lyons, J. M. 1973. Chilling injury in plants. *Ann. Rev. Plant Physiol.* 24:445–66.

MacDougal, D. T. 1936. Studies in tree growth by the dendrographic method. *Carnegie Inst. Wash. Publ.* 462.

Magness, J. R., E. S. Degman, and J. R. Furr. 1935. Soil moisture and irrigation investigations in eastern apple orchards. *U.S. Dept. Agr. Tech. Bull.* 491.

Marshall, P. E., and T. T. Kozlowski. 1974a. The role of cotyledons in growth and development of woody angiosperms. *Can. J. Bot.* 52:239–45.

———. 1974b. Photosynthetic activity of cotyledons and foliage leaves of young angiosperm seedlings. *Can. J. Bot.* 52:2023–32.

———. 1975. Changes in mineral contents of cotyledons and young seedlings of woody angiosperms. *Can. J. Bot.* 53:2021–31.

———. 1976a. Importance of photosynthetic cotyledons for early growth of woody angiosperms. *Physiol. Plant.* 37:336–40.

———. 1976b. Importance of endosperm for nutrition of *Fraxinus pennsylvanica* seedlings. *J. Exptl. Bot.* 98:572–74.

———. 1976c. Compositional changes in cotyledons of woody angiosperms. *Can. J. Bot.* 54:2473–77.

———. 1977. Changes in structure and function of epigeous cotyledons of woody angiosperms during early seedling growth. *Can. J. Bot.* 55:208–15.

McAlpine, R. G. 1961. Yellow-poplar seedlings intolerant to flooding. *J. Forestry* 59:566–68.

Mikola, P. 1962. Temperature and tree growth near the northern timber line. In *Tree Growth,* ed. T. T. Kozlowski, pp. 265–74. New York: Ronald Press.

Miller, P. R., and J. R. McBride. 1975. Effects of air pollution in forests. In *Responses of Plants to Air Pollution,* eds. J. B. Mudd and T. T. Kozlowski, pp. 196–235. New York: Academic Press.

Mudd, J. B., and T. T. Kozlowski, eds. 1975. *Responses of Plants to Air Pollution.* New York: Academic Press.

Mullin, R. E. 1971. Some effects of root dipping, root exposure, and extended planting dates with white spruce. *Forestry Chron.* 47:90–93.

Myers, C. A. 1963. Vertical distribution of annual increment in thinned ponderosa pine. *Forest Sci.* 9:394–404.

Owston, P., and T. T. Kozlowski. 1978. Growth and cold hardiness of container-grown seedlings in simulated greenhouse regimes. (Unpublished data).

Parker, J. 1952. Desiccation in conifer leaves. Anatomical changes and determination of the lethal point. *Bot. Gaz.* 114:189–98.

Parsons, R. F. 1969. Physiological and ecological tolerances of *Eucalyptus incrassata* and *E. socialis* to edaphic factors. *Ecology* 50:386–90.

Pereira, J. S., and T. T. Kozlowski. 1976a. Diurnal and seasonal changes in water balance of *Abies balsamea* and *Pinus resinosa. Oecol. Plant.* 11:397–412.

———. 1976b. Leaf anatomy and water relations of *Eucalyptus camaldulensis* and *E. globulus* seedlings. *Can. J. Bot.* 54:2868–80.

———. 1977. Influence of light intensity, temperature, and leaf area on stomatal aperture and water potential of woody plants. *Can. J. For. Res.* 7:145–53.

———. 1978. Diurnal and seasonal changes in water balance of *Acer saccharum* and *Betula papyrifera. Physiol. Plant.* 43:19–30.

Pharis, R. P., and P. J. Kramer. 1964. The effects of nitrogen and drought on loblolly pine seedlings, 1: Growth and composition. *Forest Sci.* 10:143–50.

Phillips, S. O., J. M. Skelly, and H. E. Burkhart. 1977a. Growth fluctuations of loblolly pine due to periodic air pollution levels: Interaction of rainfall and age. *Phytopathology* 67:716-20.

———. 1977b. Eastern white pine exhibits growth retardation by fluctuating air pollution levels: Interaction of rainfall, age and symptom expression. *Phytopathology* 67:721-25.

Pool, R. J. 1913. Some effects of drought on vegetation. *Science* 38:822-25.

Reed, H. S., and E. T. Bartholomew. 1930. The effects of desiccating winds on citrus trees. *Calif. Agr. Expt. Sta. Bull.* 484:1-59.

Rice, E. L. 1974. *Allelopathy.* New York: Academic Press.

Richards, P. W. 1952. *The Tropical Rain Forest.* London and New York: Cambridge Univ. Press.

Rook, D. A. 1969. Water relations of wrenched and unwrenched *Pinus radiata* seedlings on being transplanted into conditions of water stress. *New Zealand J. Forestry* 14:50-58.

———. 1971. Effect of undercutting and wrenching on growth of *Pinus radiata* D. Don seedlings. *J. Appl. Ecol.* 8:477-90.

Rowe, J. S. 1964. Environmental preconditioning with special reference to forestry. *Ecology* 45:399-403

Rowe, R. N., and D. V. Beardsell. 1973. Waterlogging of fruit trees. *Hort. Abstr.* 43:534-48.

Rudolph, T. D. 1964. Lammas growth and prolepsis in jack pine in the Lake States. *Forest Sci. Monogr.* 6.

Sakai, A., and C. J. Weiser. 1973. Freezing resistance of trees in North America with reference to tree regions. *Ecology* 54:118-26.

Sasaki, S., and T. T. Kozlowski. 1968a. The role of cotyledons in early development of pine seedlings. *Can. J. Bot.* 46:1173-83.

———. 1968b. Effects of herbicides on seed germination and early seedling development of *Pinus resinosa. Bot. Gaz.* 129:238-46.

———. 1968c. Effects of herbicides on respiration of red pine (*Pinus resinosa* Ait.) seedlings. I. S-triazine and chlorophenoxy acid herbicides. *Advan. Frontiers Plant Sci.* 22:187–202.

———. 1968d. Effects of herbicides on respiration of red pine (*Pinus resinosa* Ait.) seedlings, 2: Monuron, diuron, DCPA, dalapon, CDEC, CDAA, EPTC, and NPA. *Bot. Gaz.* 129:286–93.

———. 1969. Utilization of seed reserves and currently produced photosynthates of embryonic tissues of pine seedlings. *Ann. Bot.* 33:472–82.

———. 1970. Effects of cotyledon and hypocotyl photosynthesis on growth of young pine seedlings. *New Phytol.* 69:493–500.

Seaton, F. A. et al. 1970. *Report of the President's Advisory Panel on Timber and the Environment.* Washington, D.C.: U.S. Govt. Printing Office, 540 pp.

Siwecki, R., and T. T. Kozlowski. 1973. Leaf anatomy and water relations of six *Populus* clones. *Arboretum Kornickie* 18:83–105.

Slatyer, R. O. 1957. The influence of progressive increases in total soil moisture stress on transpiration, growth, and internal water relationships of plants. *Austr. J. Biol. Sci.* 10:320–36.

———. 1967. *Plant-Water Relationships.* London: Academic Press.

———, and J. F. Bierhuizen. 1964. Transpiration from cotton leaves under a range of environmental conditions in relation to internal and external diffusive resistances. *Austr. J. Biol. Sci.* 115–30.

Smith, K. A., and R. S. Russell. 1969. Occurrence of ethylene and its significance in anaerobic soil. *Nature* 222: 769–71.

Smithberg, M. H., and C. J. Weiser. 1968. Patterns of variation among climatic races of red-osier dogwood. *Ecology* 49:495–505.

Solberg, R. A., and D. F. Adams. 1956. Histological responses of some plant leaves to hydrogen flouride and sulfur dioxide. *Amer. J. Bot.* 43:755–60.

Spurr, S. H. 1964. *Forest Ecology.* New York: Ronald Press.

———. 1976. *American Forest Policy in Development.* Seattle: Univ. of Washington Press.

Squillace, A. E., and R. R. Silen. 1962. Racial variation in ponderosa pine. *Forest Sci. Monogr.* 2.

Stewart, R. E. 1975. Allelopathic potential of western bracken. *J. Chem. Ecol.* 1:161–69.

Stone, E. C. 1957. Dew as an ecological factor, 1: A review of the literature. *Ecology* 38:407–13.

———, and J. L. Jenkinson. 1971. Physiological grading of ponderosa pine nursery stock. *J. Forestry* 69:31–33.

Thomas, M. D. 1961. Effects of air pollution on plants. In *Air Pollution,* Monograph No. 46, pp. 233–78. Geneva: World Health Org.

Vaadia, Y., and Y. Waisel. 1963. Water absorption of the aerial organs of plants. *Physiol. Plant.* 16:44–51.

Waisel, Y. 1958. Dew absorption by plants of arid zones. *Bull. Res. Counc. Israel* 60:180–86.

———. 1960. Ecological studies on *Tamarix aphylla* (L.) Karst., 2: The water economy. *Phyton* (Buenos Aires) 15:19–28.

Wakeley, P. F. 1954. Planting the southern pines. *U.S. Forest Service, Agr. Monogr.* 18. Washington, D. C.

Wareing, P. F. 1974. Plant hormones and crop growth. *J. Roy. Soc. Arts,* Nov.: 818–27.

Waring, R. H., and S. W. Running. 1976. Water uptake, storage, and transpiration by conifers: A physiological model. In *Water and Plant Life,* eds. O. L. Lange, L. Kappen, and E.-D. Schulze, pp. 189–202. Berlin, Heidelberg, New York: Springer-Verlag.

Weinstein, L. 1975. Dose-response relationships. In *Air Pollution and Metropolitan Woody Vegetation,* eds. W. H. Smith and L. S. Dochinger, Pinchot Inst. Paper

PIEFR-PA-1. Upper Darby, Pa.: USDA Forest Service.
Weiser, C. J. 1970. Cold resistance and injury in woody plants. *Science* 169:1269–78.
Went, F. W. 1955. The ecology of desert plants. *Sci. Amer.* 192:68–76.
Whittaker, R. H. 1970. The biochemical ecology of higher plants. In *Chemical Ecology,* eds. E. Sondheimer and J. B. Simeone, pp. 43–70. New York: Academic Press.
———, and P. P. Feeny. 1971. Allelochemics: Chemical interactions between species. *Science* 171:757–70.
Wilde, S. A. 1958. *Forest Soils.* New York: Ronald Press.
———, and G. K. Voigt. 1948. Specific gravity of the wood of jack pine seedlings raised under different levels of soil fertility. *J. Forestry* 46:521–23.
Winget, C. H., and T. T. Kozlowski. 1965. Yellow birch germination and seedling growth. *Forest Sci.* 11:386–92.
———, T. T. Kozlowski, and J. E. Kuntz. 1963. Effects of herbicides on red pine nursery stock. *Weeds* 11:87–90.
Woods, D. B., and N. C. Turner. 1971. Stomatal response to changing light by four tree species of varying shade tolerances. *New Phytol.* 70:77–84.
Woods, F. W. 1959. Slash pine roots start growth soon after planting. *J. Forestry* 57:209.
Woodwell, G. M. 1970. Effects of air pollution on the structure and physiology of ecosystems. *Science* 168:429–33.
Wright, J. W. 1976. *An Introduction to Forest Genetics.* New York: Academic Press.
Wu, C. C., and T. T. Kozlowski. 1972. Some histological effects of direct contact of *Pinus resinosa* seeds and young seedlings with 2,4,5-T. *Weed Res.* 12:229–33.
Wuenscher, J. E., and T. T. Kozlowski. 1971a. The response of transpiration resistance to leaf temperature as a desiccation resistance mechanism in tree seedlings. *Physiol. Plant.* 24:254–59.

———, and T. T. Kozlowski. 1971b. Relationship of gas exchange resistance to tree seedling ecology. *Ecology* 52:1016–23.

Zahner, R. 1958. September rains bring growth gains. *Southern Forest Expt. Sta. Forest,* note 113, U.S. Forest Serv.

———. 1968. Water deficits and growth of trees. In *Water Deficits and Plant Growth,* vol. 2, ed. T. T. Kozlowski, pp. 191–254. New York: Academic Press.

———, and W. W. Oliver. 1962. The influence of thinning and pruning on the date of summerwood initiation in red and jack pines. *Forest Sci.* 8:51–63.

———, and A. R. Stage. 1966. A procedure for calculating daily moisture stress and its utility in regressions of tree growth on weather. *Ecology* 47:64–74.

———, and F. W. Whitmore. 1960. Early growth of radically thinned loblolly pine. *J. Forestry* 58:628–34.

Zavitkovski, J., and W. K. Ferrell. 1970. Effect of drought upon rates of photosynthesis, respiration, and transpiration of seedlings of two ecotypes of Douglas-fir, 2: Two-year-old seedlings. *Photosynthetica* 4:58–67.

Zelawski, W., and I. Goral. 1966. Seasonal changes in the photosynthetic rate of Scots pine seedlings grown from seed of various provenances. *Acta Soc. Bot. Pol.* 35:587–98.

THEODORE T. KOZLOWSKI, currently A. J. Riker Professor of Forestry and Director of the Biotron at the University of Wisconsin at Madison, began his long and distinguised career in the field of forestry at Syracuse University, where he received a B.S. degree in 1939. He received an M.A. degree in 1941 from Duke University, and after serving in the U.S. Army Air Force for four years, continued his studies, receiving his doctorate from Duke in 1947. He was elected to Sigma Xi, Phi Kappa Phi, and Phi Sigma honorary societies. He began his career in research and teaching at the University of Massachusetts in the department of botany, where he stayed until 1958, achieving in the meantime the rank of full professor and serving as head of the department. The following year, Dr. Kozlowski accepted a professorship at the University of Wisconsin at Madison.

As a forest biologist, Dr. Kozlowski has lectured and taught at a number of major American and foreign universities. In connection with his research he has traveled extensively and has served a consultant to various organizations, including the National Science Foundation, the National Park Service, and the Stanford Research Institute, as well as several commercial agencies. He has been Senior Fulbright Research Scholar at Oxford University, Visiting Scientist at the Society of American Foresters, and Visiting Biologist at the American Institute of Biological Sciences. Dr. Kozlowski has also been the recipient of a number of distinguished awards, including the Barrington Moore Memorial Award, given by the Society of American Foresters for "outstanding achievement in biological research contributing to the advancement of forestry." He has been a member and officer of a number of professional organizations as well. In 1978 he was awarded the honorary degree of Doctor of Science, Honoris Causa, by De L'Universite Catholique de Louvain in Belgium.

In addition to his many other achievements, Dr. Kozlowski has authored or served as co-author or editor for and extensive list of titles, all important contributions to the literature in the area of plant physiology. He also served on the editorial boards of the periodicals *Forest Science* and *Ecology* and has been the associate editor of the *Canadian Journal of Forest Research*.